THE DEVELOPMENT OF DRIs 1994–2004

Lessons Learned and New Challenges

Workshop Summary

AF754801

Marla Sheffer and Christine Lewis Taylor, *Rapporteurs*

Food and Nutrition Board

INSTITUTE OF MEDICINE
OF THE NATIONAL ACADEMIES

THE NATIONAL ACADEMIES PRESS
Washington, D.C.
www.nap.edu

THE NATIONAL ACADEMIES PRESS 500 Fifth Street, N.W. Washington, DC 20001

NOTICE: The project that is the subject of this report was approved by the Governing Board of the National Research Council, whose members are drawn from the councils of the National Academy of Sciences, the National Academy of Engineering, and the Institute of Medicine.

This study was supported by Contract No. HHSP2332004509XI TO #8 between the National Academy of Sciences and U.S. Department of Health and Human Services; Contract No. AG-0202-P-06-0014 between the National Academy of Sciences and U.S. Department of Agriculture; and Contract No. 4500164550 between the National Academy of Sciences and Health Canada. Additional support came from the Canadian Institutes of Health Research. Any opinions, findings, conclusions, or recommendations expressed in this publication are those of the author(s) and do not necessarily reflect the view of the organizations or agencies that provided support for this project.

International Standard Book Number-13: 978-0-309-11562-9
International Standard Book Number-10: 0-309-11562-0

Additional copies of this report are available from the National Academies Press, 500 Fifth Street, N.W., Lockbox 285, Washington, DC 20055; (800) 624-6242 or (202) 334-3313 (in the Washington metropolitan area); Internet, http://www.nap.edu.

For more information about the Institute of Medicine, visit the IOM home page at: **www.iom.edu.**

Printed in the United States of America.

The serpent has been a symbol of long life, healing, and knowledge among almost all cultures and religions since the beginning of recorded history. The serpent adopted as a logotype by the Institute of Medicine is a relief carving from ancient Greece, now held by the Staatliche Museen in Berlin.

Suggested citation: Institute of Medicine (IOM). 2008. *The development of DRIs 1994–2004: Lessons learned and new challenges—workshop summary.* Washington, DC: The National Academies Press.

"Knowing is not enough; we must apply.
Willing is not enough; we must do."

—Goethe

INSTITUTE OF MEDICINE

OF THE NATIONAL ACADEMIES

Advising the Nation. Improving Health.

THE NATIONAL ACADEMIES

Advisers to the Nation on Science, Engineering, and Medicine

The **National Academy of Sciences** is a private, nonprofit, self-perpetuating society of distinguished scholars engaged in scientific and engineering research, dedicated to the furtherance of science and technology and to their use for the general welfare. Upon the authority of the charter granted to it by the Congress in 1863, the Academy has a mandate that requires it to advise the federal government on scientific and technical matters. Dr. Ralph J. Cicerone is president of the National Academy of Sciences.

The **National Academy of Engineering** was established in 1964, under the charter of the National Academy of Sciences, as a parallel organization of outstanding engineers. It is autonomous in its administration and in the selection of its members, sharing with the National Academy of Sciences the responsibility for advising the federal government. The National Academy of Engineering also sponsors engineering programs aimed at meeting national needs, encourages education and research, and recognizes the superior achievements of engineers. Dr. Charles M. Vest is president of the National Academy of Engineering.

The **Institute of Medicine** was established in 1970 by the National Academy of Sciences to secure the services of eminent members of appropriate professions in the examination of policy matters pertaining to the health of the public. The Institute acts under the responsibility given to the National Academy of Sciences by its congressional charter to be an adviser to the federal government and, upon its own initiative, to identify issues of medical care, research, and education. Dr. Harvey V. Fineberg is president of the Institute of Medicine.

The **National Research Council** was organized by the National Academy of Sciences in 1916 to associate the broad community of science and technology with the Academy's purposes of furthering knowledge and advising the federal government. Functioning in accordance with general policies determined by the Academy, the Council has become the principal operating agency of both the National Academy of Sciences and the National Academy of Engineering in providing services to the government, the public, and the scientific and engineering communities. The Council is administered jointly by both Academies and the Institute of Medicine. Dr. Ralph J. Cicerone and Dr. Charles M. Vest are chair and vice chair, respectively, of the National Research Council.

www.national-academies.org

PLANNING COMMITTEE FOR DIETARY REFERENCE INTAKES REVIEW WORKSHOP

JOHN W. SUTTIE (*Chair*), Department of Biochemistry, University of Wisconsin, Madison
STEPHANIE A. ATKINSON, Nutrition and Metabolism Research Laboratory, Department of Pediatrics, McMaster University, Hamilton, Ontario
JANET C. KING, University of California at Berkeley and Davis, Children's Hospital Oakland Research Institute, Oakland
SUSAN TAYLOR MAYNE, Division of Chronic Disease Epidemiology, Yale School of Public Health, New Haven, CT
HUGUETTE TURGEON O'BRIEN, Department of Food Sciences and Nutrition, Laval University, Quebec City, Quebec
WILLIAM M. RAND, Department of Public Health and Family Medicine, Tufts University School of Medicine, Boston, MA
ROBERT M. RUSSELL, Jean Mayer USDA Human Nutrition Research Center on Aging, Tufts University, Boston, MA

Staff

CHRISTINE LEWIS TAYLOR, Study Director and Scholar
HEATHER DEL VALLE, Research Associate
SANDRA AMAMOO-KAKRA, Program Associate
MARLA SHEFFER, *Consultant Writer*, Orleans, Ontario

*IOM boards do not review or approve individual reports and are not asked to endorse conclusions and recommendations. The responsibility for the content of the summary rests with the rapporteurs and the institution.

Reviewers

This report has been reviewed in draft form by individuals chosen for their diverse perspectives and technical expertise, in accordance with procedures approved by the National Research Council's Report Review Committee. The purpose of this independent review is to provide candid and critical comments that will assist the institution in making its published report as sound as possible and to ensure that the report meets institutional standards for objectivity, evidence, and responsiveness to the study charge. The review comments and draft manuscript remain confidential to protect the integrity of the deliberative process. We wish to thank the following individuals for their review of this report:

Frances Seligson, Hershey, PA
John W. Suttie, Department of Biochemistry, University of Wisconsin
John E. Vanderveen, Food and Drug Administration, Rockville, MD
Susan Whiting, College of Pharmacy and Nutrition, University of Saskatchewan, Canada

Although the reviewers listed above have provided many constructive comments and suggestions, they were not asked to endorse the final draft of the report before its release. The review of this report was overseen by **Hugh Tilson, M.D., Dr.P.H., M.P.H.,** University of North Carolina, who was appointed by the Institute of Medicine. He was responsible for making certain that an independent examination of this report was carried out in accordance with institutional procedures and that all review comments were carefully considered. Responsibility for the final content of this report rests entirely with the author and the institution.

Contents

Acronyms

AI	Adequate Intake
ALA	α-linolenic acid
AMDR	Acceptable Macronutrient Distribution Range
AROI	Acceptable Range of Oral Intake
ATBC	Alpha-Tocopherol Beta-Carotene
BMI	body mass index
BMR	basal metabolic rate
BW	body weight
CDC	Centers for Disease Control and Prevention
CHD	coronary heart disease
CV	coefficient of variation
CVD	cardiovascular disease
DHA	docosahexaenoic acid
DNA	deoxyribonucleic acid
DRI	Dietary Reference Intake
EAR	Estimated Average Requirement
EC	European Commission
EER	Estimated Energy Requirement
EFSA	European Food Safety Authority
EPA	eicosapentaenoic acid
EU	European Union

FACA	Federal Advisory Committee Act
FDA	U.S. Food and Drug Administration
FFQ	food frequency questionnaire
FNB	Food and Nutrition Board
IOM	Institute of Medicine
ISU	Iowa State University
LDL	low-density lipoprotein
LOAEL	lowest-observed-adverse-effect level
NAS	National Academy of Sciences
NCI	National Cancer Institute
NHANES	National Health and Nutrition Examination Survey
NIH	National Institutes of Health
NOAEL	no-observed-adverse-effect level
NRC	National Research Council
OPEN	Observing Protein and Energy Nutrition
PICO	population, intervention, comparator, outcome
RAE	retinol activity equivalent
RDA	Recommended Dietary Allowance
RNI	Recommended Nutrient Intake
SD	standard deviation
SEBR	systematic evidence-based review (also abbreviated in some documents as SR)
TSH	thyroid stimulating hormone
UF	uncertainty factor
UK	United Kingdom
UL	tolerable upper intake level
USDA	U.S. Department of Agriculture
WHO	World Health Organization
WIC	Special Supplemental Nutrition Program for Women, Infants and Children

Preface

In what ways can the process for developing Dietary Reference Intakes (DRIs) be enhanced? The workshop entitled "The Development of DRIs 1994–2004: Lessons Learned and New Challenges" offered a valuable window into the issues and challenges inherent in the development of nutrient reference values. The dialogue—carried out under the auspices of the Institute of Medicine (IOM), Food and Nutrition Board (hereafter referred to jointly as the IOM)—was enriched by the 10 years of experience in deriving the expanded set of values known as the DRIs, plus the decades of experience that grounded the earlier Recommended Dietary Allowances for the United States and the Recommended Nutrient Intakes for Canada. The lessons learned and the knowledge gained will guide decisions about the next phase of the DRIs. To paraphrase one participant, we are now asking better questions.

In 2006, the IOM, with support from the United States and Canadian governments, undertook an effort to synthesize the research needs identified during the 10 years of DRI development. While the workshop summarized here was predicated on the fact that the development of DRIs is improved by better data, its focus was different. Its goals were to examine the framework[1] and conceptual underpinnings for developing DRIs and to identify issues important for enhancing the process of DRI development.

[1]Agreement is lacking as to whether the preferred or appropriate term is "framework," "approach," "model," or "paradigm." The term generally used in this summary is "framework." However, given the lack of consensus, no efforts were made to change speakers' or discussants' remarks to universally refer to a single term.

The United States and Canadian governments again served as sponsors for this important effort.

The process for developing nutrient reference values in the United States and Canada has evolved over time. For example, the earliest editions undertaken by the National Research Council, dating back to the 1940s and 1950s, included fewer nutrients and less background than did later editions. Beginning in 1994, efforts overseen by the IOM resulted in DRIs that specified several kinds of reference values and provided more volumes of accompanying explanation as well as guidance for users. These changes, of course, were a function of the increasing knowledge base in the field of nutrient requirements as well as evidence of the consequences of excessive intake. However, they were also due to the increasing interest in providing transparency for the decision-making process and in communicating better with those responsible for making the public health policy that depends on the reference values.

The workshop was designed to use the existing framework for DRI development as a basis for the discussions and to consider the components of the framework in sequence. Consideration of the pros and cons of the current conceptual underpinnings of the framework opened the workshop, followed by the general "road map" for decision making and the needed scientific criteria. Next, the challenges associated with providing guidance for users were explored. The workshop concluded with an array of issues germane to the future process for developing DRIs, including strategies for updating and revising existing DRIs and opportunities for stakeholder input.

Many topics were interrelated and common themes often emerged during different discussions. Transparency and the need for more information on the rationale behind the decisions made were important themes sounded during the workshop. There was also interest in determining a method whereby the uncertainty surrounding the reference values could be better articulated and made known to users. In looking to the future, the advantages of focusing on single nutrients or groups of similar nutrients were highlighted. Appendix C of this summary contains a brief listing of reoccurring workshop discussions and may be useful to readers.

Given the history that many participants brought with them to the workshop, it was not surprising that the discussions were rich and focused. Nor was it surprising that the numerous successes of the DRI process were readily acknowledged. What was remarkable and very gratifying was the willingness of those close to the DRI process to openly discuss newer options and to readily acknowledge the appropriateness of some changes.

Although many presenters, discussants, and panelists expressed viewpoints and recommended specific strategies, their perspectives and recommendations should not be viewed as workshop conclusions or recommendations. The workshop was designed to identify issues and foster discussion, not to identify consensus recommendations. The discussions will be useful in planning the next stage of DRI activities, and clearly such planning activities will benefit greatly from the workshop conversations.

Christine Lewis Taylor, IOM Scholar and Study Director
Linda D. Meyers, Director, Food and Nutrition Board

1

Workshop Introduction[1]

Dr. John Suttie, chair of the workshop, welcomed participants on behalf of the Food and Nutrition Board (FNB), Institute of Medicine (IOM).[2] He also gave a brief overview of the Dietary Reference Intakes (DRIs) and the format of the workshop. He then introduced Dr. Christine Taylor, Study Director and IOM Scholar, who described the current DRI framework and the issues that have been raised about the DRI development process. Dr. Suttie then invited Dr. Dennis Bier to offer his welcome as FNB chair.

WELCOME AND OPENING REMARKS

Presenter: John Suttie

This workshop, through the generous support of the U.S. and Canadian governments, focuses on the process by which the DRIs were developed. It has been designed to identify lessons learned and to offer opportunities to discuss the ways in which the process might be enhanced.

The development of reference values for nutrients has a long history. In 1994 the IOM, with the guidance of the FNB, undertook activities that resulted in major changes in how reference values were developed and

---

The planning committee's role was limited to planning the workshop, and the workshop summary has been prepared by the workshop rapporteur (with acknowledgment of the assistance of staff as appropriate) as a factual summary of what occurred at the workshop.

[1]This chapter is an edited version of remarks presented by Drs. Suttie and Taylor at the workshop.

[2]FNB, IOM hereafter jointly referred to as IOM.

presented, ultimately leading to the new DRIs. In many ways, these efforts considerably advanced the approach used to develop nutrient reference values. At that time we recognized that after some experience had been gained using this new approach, it would be worthwhile to pause and examine not only our successes, but the ways in which the approach could be improved. We are now at that point.

Background for Workshop Discussions

Reference values known in the United States as Recommended Dietary Allowances (RDAs) and in Canada as Recommended Nutrient Intakes (RNIs) were used through the 1990s. They were established primarily to set nutrition and health policy. In 1994, in response to important changes in the nutrition field as well as the recognition that for many nutrients the single RDA or RNI values did not meet the expanding needs for nutrient reference values, the IOM began an initiative to develop a new, broader set of values known as the DRIs. The U.S. and Canadian governments supported this initiative.

More specifically, the DRIs as reference values now

- include upper levels of intake, where appropriate;
- incorporate chronic disease endpoints within the array of endpoints that may serve to establish adequate intake or upper intake levels;
- include "non-classical" nutrients;
- specifically highlight concepts of probability and risk for defining reference values; and
- are associated with publications intended to guide users of DRIs.

The DRI component values are shown in Box 1-1. They are described and contained in six volumes published by the IOM between 1997 and 2005. To help users understand the DRIs, given the expansion of both the nutrient reference value approach and the types of reference values issued, two publications were created to provide general guidance for users, one focused on planning and the other on assessment. In 2006, the IOM issued *Dietary Reference Intakes: The Essential Guide to Nutrient Requirements*, which is available in English and French.

Workshop Development

A planning committee was convened early this year to assist the IOM in formulating the content and format of the workshop and in identifying candidates to serve as speakers, discussants, and panel members. The committee specified background materials to help participants prepare for the

BOX 1-1
Current Dietary Reference Intake (DRI) Components

Estimated Average Requirement (EAR): Reflects the estimated median requirement and is particularly appropriate for applications related to planning and assessing intakes for groups of persons.

Recommended Dietary Allowance (RDA): Derived from the EAR and covers the requirements for 97 percent of the population.

Tolerable Upper Intake Level (UL): Highest average intake that is likely to pose no risk.

Adequate Intake (AI): Used when an EAR/RDA cannot be developed; average intake level based on observed or experimental intakes.

Acceptable Macronutrient Distribution Range (AMDR): An intake range for an energy source associated with reduced risk of chronic disease.

workshop. These materials were available on the IOM website for review and comment prior to the workshop.

As outlined on the agenda (see Appendix A), there are four major topic areas: (1) the conceptual framework for DRI development, (2) criteria for scientific decision making, (3) general guidance for users of DRIs, and (4) the future process for DRI development. In order to foster diverse discussions and include a range of perspectives, the workshop is organized around a series of presentations that are complemented by topic-designated discussions as well as broader panel discussions. Considerable time has been set aside for audience members to provide their views on the DRI process.

In summary, this workshop offers a unique opportunity to consider the DRI process and raise issues important to its enhancement. Our intent is to have an open discussion that will prove useful as we consider the next steps for DRI development. Although the workshop is not charged with coming to closure on the issues raised and will not conclude with consensus recommendations, it should provide a useful spectrum of stakeholder comments on this important activity.

OVERVIEW OF CURRENT DRI FRAMEWORK AND ISSUES RAISED

Presenter: Christine Taylor

A new approach to nutrient reference values was put in place in the mid-1990s to respond to the expanded uses of the values and to the newer understandings of the role of nutrients. With 10 years of experience behind us, we now have the basis for considering how the approach has performed and whether enhancements or modifications are needed. This presentation sets the groundwork for the discussion by describing the current framework and identifying the issues raised.

Outlining the Framework

The original intent was for the DRI framework to be developed and "fleshed out" through the experience of developing the values. Therefore, the DRI framework does not exist in a single or codified document, but is gleaned from several sources. Principles for DRI development were articulated in a general way in the 1994 publication, *How Should the Recommended Dietary Allowances Be Revised?* (IOM, 1994). Further information can be found in the first DRI volume issued in 1997 (IOM, 1997), in the 1998 monograph that describes the approach for the upper levels of intake (IOM, 1998), and in the 2006 *Dietary Reference Intakes: The Essential Guide to Nutrient Requirements* (IOM, 2006).

Basic components of the DRI framework are shown in Figure 1-1.[3] The initial component is the set of conceptual underpinnings that may be referred to as first principles. It includes the task of problem formulation, a dialogue to ensure common understanding, and clarification of the reasons for the activity. Next is what can be referred to as a road map—principles and criteria for the study committees to use as they develop nutrient reference values. Following these activities is general guidance for users to outline appropriate strategies for applying the reference values.

Some important DRI-related activities are outside the scope of the workshop and should be addressed separately. One is the critical set of issues surrounding the research needed to elucidate the basic physiology, metabolism, and homeostatic mechanisms—in essence, the data that provide the raw materials for DRI development. Second are the science and considerations needed to allow application of the DRIs under specific situations that require in-depth study and consideration beyond that which can be provided in general guidance to users (e.g., the use of the DRIs for nutrition labeling or in developing food assistance programs under U.S. federal regulations).

The workshop agenda (Appendix A)—and hence the workshop presentations and discussions—moves sequentially through the framework as outlined. Session 1 focuses on the underpinnings and then addresses general road map considerations. Session 2 continues with the road map, focusing

[3]The workshop planning committee used a more complex schematic of the DRI process (see Appendix D) in its deliberations.

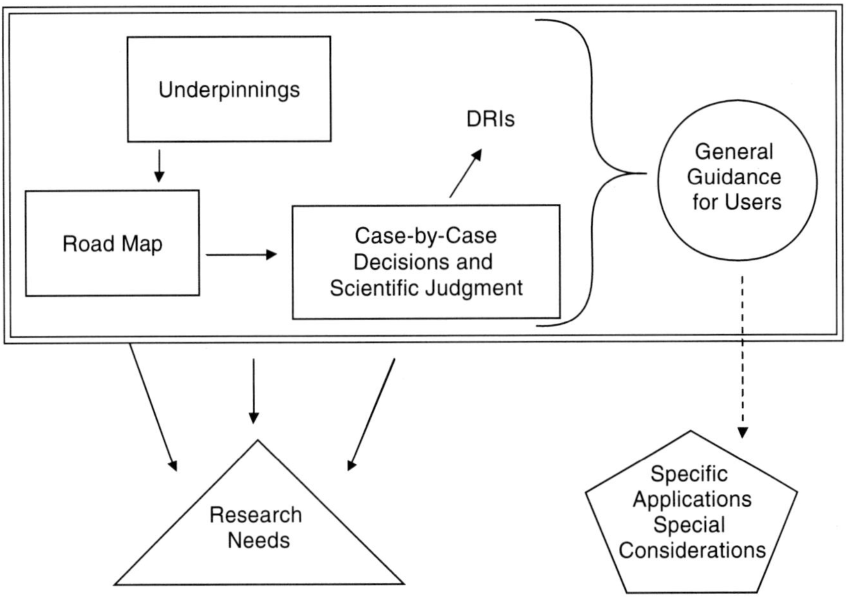

FIGURE 1-1 DRI framework schematic. Framework components that are the subject of the workshop are identified in the box; other important activities are identified outside the box. (See also Appendix D.)
NOTE: DRIs = Dietary Reference Intakes.

on six topics. Session 3 explores the general guidance for users. Session 4 considers the future process. Although the activities in the workshop are not strictly linear and a number of issues are crosscutting and iterative, the organization of the workshop provides a reasonable way to manage and focus the needed discussions.

Conceptual Underpinnings

The conceptual underpinnings have the following two major components:

- Purpose of the DRIs (i.e., the goals and guiding principles)
- Setup of DRIs (i.e., the kinds of values to be expressed, the types of endpoints that are appropriate, and the nature of the nutrient substances for DRI development)

DRIs were identified in 1994 as standards to serve as a goal for good nutrition. They were to focus on groups of healthy persons, but the desirability of exploring ways to evaluate individuals was highlighted. From 1997 to 2006, the DRIs were identified as serving the purpose of planning and assessing diets for healthy populations and other purposes.

Question: Does the development of standards for planning and assessing healthy groups and individuals remain the general purpose of the DRIs?

Turning to the DRI setup, as early as 1989 it was recognized that the RDA value alone was not sufficient to meet all the reference value needs. In 1994, DRIs were foreshadowed as a set of values that would include more than an RDA—specifically, an estimation of a median or average requirement as well as an indication of an upper level of intake. These became the Estimated Average Requirement (EAR) and the tolerable upper intake level (UL). Activities in 1997 and beyond added Adequate Intakes (AIs) and the Acceptable Macronutrient Distribution Range (AMDR).

Questions: Do the EAR, RDA, and UL continue to be desirable values? Is the AI still useful and needed, and should it be better defined? Does the AMDR pave the way to considering macronutrients using a different approach?

Also relevant to the DRI setup is the type of endpoints that are appropriate. The existing framework addresses two types: endpoints to ensure adequacy and endpoints to avoid excess, each of which include chronic and non-chronic disease endpoints. Most DRIs, whether focused on adequacy or on upper levels, are based on non-chronic disease endpoints. The incorporation of chronic disease endpoints has been challenging. In addition, some have raised the question about the desirability of providing values based on more than one endpoint for specific age/gender groups. Although different endpoints for different age/gender and life stage groups are now used as appropriate, a single endpoint, rather than multiple endpoints, is selected for each group.

Questions: Should we continue to include reduction of chronic disease risk as an endpoint option? Should we explore the option of issuing DRI values for multiple endpoints for a single group, allowing users to select their preferred endpoint?

Finally, the question arises about nutrient substances appropriate for DRI development. Historically, these substances have been essential or so-called classic nutrients. More recently, compounds found naturally in foods have been included—usually those with potential risks or possible benefits to health, such as fiber, cholesterol, and saturated fat.

> **Questions:** Should the focus of the DRIs continue to expand beyond classic nutrients? Is a modified DRI approach needed to address macronutrients and nonessential nutrient substances?

Road Map

In essence, the road map focuses on the steps of DRI development. In planning the workshop, we learned that much work needs to be done to provide a road map for the study committees. However, a set of initial questions needs to be addressed.

> **Initial Questions:** Can we provide more specific guidance on scientific decision making to help study committees clarify the concepts and tasks and to promote consistency across study committees? Can we provide guidance to study committees on the use of scientific judgment in the face of limited data that would allow the derivation of the judgment to be more transparent and better documented?

Other issues relate to the general process for developing DRIs including the ability to specify an organizing scheme and the role of systematic evidence-based reviews.

> **Process Questions:** What is the role of systematic evidence-based reviews in DRI development? Can an organizing scheme for DRI development be specified?

An apparent organizing scheme is outlined in Figure 1-2. Study committees first review the data and develop the DRI values. There is some concern as to whether study committees consistently attend to the useful next step of reviewing the exposure (or intake) assessment for the population of interest, placing the DRI values in context, and characterizing the risk of inadequate or excessive intakes. Given this concern, the discussions during the workshop are expected to focus on the agreed-upon general steps for DRI development and on enhancing the risk characterization and related "contextual discussions," particularly as they relate to clarifying uncertainty and precision of the value, and comparing the values with the current estimated intake. Another focus is increased efforts to enhance the formal collaboration of the steps for determining reference values for adequacy and for determining upper levels of intake.

Discussions about specific road map tasks related to scientific decision making can be used to identify the best questions about these important activities and the best approaches for developing better criteria for use by study committees. These road map tasks include

- the approach for selecting endpoints, notably selection criteria and consistency of the approach used;
- available methodologies for approximating dose–response relationships in the face of limited data, given the critical nature of such data;
- strategies for extrapolation, scaling, and interpolation of values from a particular age/gender group to an unstudied group;
- adjusting for data uncertainty, including the possibility of a specific rationale and criteria;
- strengths and limitations of dietary intake estimates as they may impact risk characterization; and
- important areas to be monitored during the process of DRI development as they relate to environmental, genomic, and physiological factors.

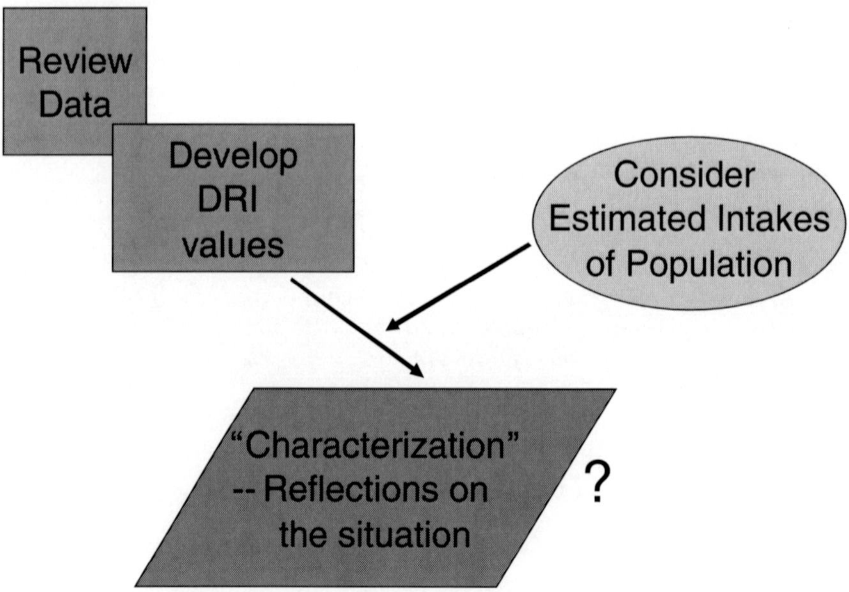

FIGURE 1-2 Basic steps in DRI development.
NOTE: DRI = Dietary Reference Intake.

General Guidance for Users

Issues raised concerning general guidance for users relate primarily to the publications on applications in dietary assessment and planning (IOM, 2000a, 2003a). The guidance outlines a general approach for using the various DRI values and is linked to the existing purpose of DRIs: planning and assessment for groups and individuals.

It is important to distinguish between "uses" as they relate generally to the purpose of DRIs and "uses" as they relate to general applications of DRIs in real-world settings when dietary planning and assessment are taking place. Examples of issues related to "uses" as they relate to the purpose of DRIs—and thus earmarked for discussion as components of the conceptual underpinnings—include the uncertainty surrounding or precision of the reference values, and how appropriate those values are for use with diseased populations.

Diverse issues have been raised about guidance for users, many of which will require in-depth and focused discussions.

> **Questions:** How should AIs be used to address planning and assessing, especially within the context of the total diet? Can clarification be provided on the differences between groups and individuals and between applications for small groups and large groups? What practical guidance and tools can be used to assist practitioners?

Looking to the Future

A pivotal aspect of discussion about the DRI development process is how it will work in the future. Key topics to be addressed include

- ways to enhance transparency of the decision-making process;
- criteria and strategies for updating and reviewing the DRIs;
- how to determine what and when "new" nutrient substances are appropriate for DRI development;
- options for stakeholder input within the DRI process; and
- issues that may emerge in the future.

With this framework as a basis, we hope the rationale for the questions posed will be clear and the workshop discussions will be well grounded. Thank you in advance for what we are certain will be a rich conversation among a diverse set of experts and stakeholders.

2

Conceptual Framework for DRI Development: Session 1[1]

Prior to the workshop, Session 1 participants were asked to consider several general questions (shown in Box 2-1) in preparing their presentations. Session 1 addressed both the conceptual underpinnings and several overarching "roadmap" issues as described in Workshop Introduction (see Chapter 1).

The session was moderated by Dr. Stephanie Atkinson of McMaster University. Dr. Robert Russell of Tufts University discussed the pros and cons of the current framework for Dietary Reference Intake (DRI) development. Two case studies were then presented. Dr. Paula Trumbo, a former study director for DRI micronutrient, macronutrient, fiber, and water and electrolyte study committees who is now at the U.S. Food and Drug Administration (FDA), explored considerations when applying the DRI framework to chronic disease endpoints. Dr. Allison Yates, who served as director of the Food and Nutrition Board (FNB) from 1994 through 2003 and is now director of the Agricultural Research Service Human Nutrition Center at the U.S. Department of Agriculture (USDA), discussed applying the DRI framework to non-chronic disease endpoints.

Perspectives on the DRIs were offered by Dr. George Beaton and Dr. Janet King. Dr. Beaton is professor emeritus at the University of Toronto and has served as a consultant to the Institute of Medicine (IOM). Dr. King

[1]This chapter is an edited version of remarks presented by Drs. Russell, Trumbo, Yates, Beaton, King, Lichtenstein, and Yetley at the workshop. Discussions are composites of input from various panel members, discussants, presenters, moderators, and audience members.

BOX 2-1
General Questions for Session 1 Participants

Conceptual Underpinnings

- How has the Dietary Reference Intake (DRI) framework "held up" over time?
- What is the general purpose of the DRIs? Is it still for planning and assessing?
- Do the Estimated Average Requirements (EARs), Recommended Dietary Allowances (RDAs) and Tolerable Upper Intake Levels (ULs) continue to be desirable values? Is the Adequate Intake (AI) useful and needed? Does the Acceptable Macronutrient Distribution Range (AMDR) pave the way to considering macronutrients using a different approach?
- Should we continue to include chronic disease risk as an endpoint option? Should we explore multiple endpoints for the same age/gender group?
- Should the focus of the DRIs continue to expand beyond classic nutrients? Is a modified DRI approach needed to address macronutrients and nonessential nutrient substances?

Overarching Road Map Issues

- What is the role of systematic evidence-based reviews (SEBRs) in DRI development?
- Can an organizing scheme for DRI development be specified?

is senior scientist at the Children's Hospital Oakland Research Institute and is a former chair of the FNB.

Dr. Alice Lichtenstein of Tufts University examined the issues in applying systematic evidence-based review (SEBR) approaches to DRI development. Dr. Elizabeth Yetley, a Senior Nutrition Research Scientist with the Office of Dietary Supplements at the National Institutes of Health, discussed whether risk assessment is a relevant organizing structure for the DRI development process.

Designated discussants followed Drs. Russell, Trumbo, and Yates, and a designated discussant engaged Drs. Lichtenstein and Yetley. In each case, the discussions were followed by input from the workshop audience. The session concluded with a panel discussion, at which point the session was again opened to the audience for comment.

CURRENT FRAMEWORK FOR DRI DEVELOPMENT: WHAT ARE THE PROS AND CONS?

Presenter: Robert M. Russell

In 1994, two major changes were made to the development of reference values. One was that the values could be based on an endpoint associated with the risk of chronic disease. The second was that reference values in addition to the Recommended Dietary Allowance (RDA) would be provided to address the increasingly broad applications of reference values. However, these major changes to the DRI development process have both pros and cons.

Reference Values Expressed: EARs, RDAs, and AIs

The Estimated Average Requirement (EAR) is the level of intake for which the risk of inadequacy would be 50 percent. The RDA is two standard deviations (SDs) above the EAR, covering 97 percent of the population.

The Adequate Intake (AI) as a reference value was not envisioned until the lack of dose–response data precluded study committees from determining the level at which the risk of inadequacy would be 50 percent. This was often exacerbated by a lack of longitudinal studies. As a result, AIs were generally set when an EAR could not be established.[2] These include calcium, vitamin D, chloride, chromium, fluoride, potassium, manganese, sodium, and vitamin K.

For calcium, an AI was issued due to uncertainty about methods used in older balance studies, a lack of concordance between observational and experimental data (i.e., the mean intakes of the population are lower than the values needed to achieve calcium retention), and a lack of longitudinal dose–response data to verify an association between the amounts needed for calcium retention and bone fracture or bone loss.

For vitamin D, an AI was developed because the study committee did not know how much dietary vitamin D is needed to maintain normal calcium metabolism and bone health, primarily because vitamin D is a complicated hormone: Exposure to sunlight, skin pigmentation, the latitude at which one lives, and the amount of clothing one wears all affect the amount of vitamin D needed. Furthermore, there were uncertainties

[2]An exception is the reference value for young infants, for whom AIs were specifically determined as opposed to developed when an EAR could not be developed. The AI for young infants has generally been the average intake by full-term infants born to healthy, well-nourished mothers and exclusively fed human milk. The only exception to this criterion is vitamin D, which occurs in low concentrations in human milk (IOM, 2006).

about the accuracy of the vitamin D food composition database and levels of food fortification.

When a chronic disease endpoint was selected as the basis for a reference value—which occurred for five nutrients—all of the reference values were AIs rather than EARs. Calcium and vitamin D AIs were set primarily on the basis of experimental data on bone density and fracture, fluoride on dental caries, potassium on hypertension, and fiber on coronary artery disease.

The selection of an endpoint for EARs presented some difficulties. A variety of endpoints were used. For example, maximum glutathione reductase activity was the endpoint used for selenium. A factorial approach[3] was used for vitamin A, zinc, and iron. The maximum neutrophil concentration that would give minimal urinary loss was used to determine the vitamin C EAR. Physiological function was used for vitamin E (the level that would inhibit peroxide-induced hemolysis) and vitamin B_{12} (maintaining a normal hematological status).

The study committees encountered numerous data gaps. The prime one was the lack of defined health-related endpoints associated with status and a lack of biomarkers to define chronic disease. Age-specific data were lacking, so extrapolation was used. Also, there was a lack of information on variability of responses (needed to calculate RDAs). As already mentioned, another data gap was the lack of dose–response data (ending up with AIs) combined with a lack of long-term studies. Adding to this list are the lack of knowledge as to which systems dysfunction with excess, as seen with bone, and the lack of uniform rules on how to apply uncertainty factors.

Another problem has been extrapolation. Using the case of vitamin A, the AI for 0- to 6-month-olds is 400 μg retinol activity equivalents (RAEs) per day. The study committee extrapolated up for the 7- to 12-month-olds to get an AI of 500 μg RAE/day, which is very close to the tolerable upper intake level (UL) (based on bulging fontanels) of 600 μg RAE/day. In using these numbers, more than half the infants (4–5 months old) in the USDA's Special Supplemental Nutrition Program for Women, Infants and Children (WIC) are eating above the UL, yet adverse effects on these infants have not been observed. Another odd observation is the lower requirement for 1- to 3-year-olds (300 μg RAE/day) than for 7- to 12-month-olds (500 μg RAE/day), because the AI for 7- to 12-month-olds was extrapolated up from 0- to 6-month-olds and the EAR for 1- to 3-year-olds was extrapolated down from the adult number. The validity of these numbers is therefore questionable.

[3]A factorial approach can take several forms but generally derives a total nutrient requirement by summing the individual physiological needs of various functional components (e.g., body maintenance, milk synthesis, skin sloughing).

The major challenge in deriving an RDA from an EAR is variance. To establish an RDA, one determines the EAR, assesses the variability, then calculates the RDA as the EAR plus two SDs. However, variance is not known for most nutrients, and a coefficient of variation (CV) is assumed instead. A 10 percent CV was assumed for thiamin, riboflavin, niacin, vitamin B_6, folate, vitamin B_{12}, vitamin C, vitamin E, selenium, and zinc.

The CV is known for some nutrients, such as vitamin A. Although the study committee was initially enthusiastic about using a physiological endpoint (abnormal dark adaptation) for determining an EAR for vitamin A, the pooled data from four studies gave a CV of 40 percent. Therefore, the study committee decided not to use dark adaptation as the endpoint, and no EAR or RDA was established on this basis. Instead, a higher EAR (625 µg for men and 500 µg for women, compared with 300 µg) was determined using a factorial approach.

Reference Values Expressed: ULs

The UL is the highest level of daily nutrient intake that poses no risk of an adverse effect to almost any individual in the general population. It is not a recommended or desirable level of intake. It is derived by dividing a no-observed-adverse-effect level (NOAEL) or a lowest-observed-adverse-effect level (LOAEL) by an uncertainty factor.

A concern is that the uncertainty factor is subjective. The sources of uncertainty that the study committees considered were interindividual variation, extrapolation from animals to humans, short-term versus chronic exposures, use of a LOAEL instead of a NOAEL, small numbers of people studied, and the severity of the effects (the higher the severity, the higher the uncertainty factor). The example in Box 2-2 illustrates the subjectivity that study committee members face in trying to derive logical and scientifically valid numbers.

Applicability of the Framework to All Nutrient Substances

The framework did not "fit" well for establishing reference values for fat and macronutrients. Such substances are not essential and have no beneficial role, except for essential fatty acids and amino acids. Rather, an Acceptable Macronutrient Distribution Range (AMDR) for fat was determined to be 20–35 percent of calories. Furthermore, a UL was not provided for the effects of intakes of saturated fat or *trans* fat on low-density lipoprotein (LDL) cholesterol, as coronary heart disease (CHD) risk increases progressively. For fiber, an AI was set on the basis of heart disease prevention, as the effect on CHD occurs continuously across the range of intakes. No UL could be determined for fiber, because fiber intake is accompanied

BOX 2-2
Vitamin A and Uncertainty Factors

Four adverse effects were considered in setting a tolerable upper intake level (UL) for vitamin A: bone mineral density, liver toxicity, teratogenicity (for women of reproductive age), and bulging fontanels (for infants). In a study of the daily dietary intake of retinol associated with risk for hip fracture in two populations in Sweden and the United States (Melhus et al., 1998), it was determined that there was a rise in the risk for hip fracture above a vitamin A intake of 1,500 μg/day. However, two other papers were unable to show any effect of vitamin A intake on bone mineral density (Sowers and Wallace, 1990; Ballew et al., 2001). Therefore, the United States decided to use liver toxicity as the critical effect for the general adult population and derived a UL of 3,000 μg/day (twice as high as the UL based on hip fracture). For women of reproductive age, a UL of 3,000 μg/day for teratogenicity was determined based primarily on a study by Rothman et al. (1995).

The United Kingdom (UK) panel decided that the Rothman et al. (1995) paper was biased and did not set any UL for teratogenicity, as it considered the evidence base inadequate. It suggested that intakes greater than 1,500 μg/day may be inappropriate and advised pregnant women not to take vitamin A supplements.

The European Union (EU), looking at the same database used by the Institute of Medicine (IOM) study committee and the UK panel, established a UL of 3,000 μg/day, the lowest-observed-adverse-effect level (LOAEL) for teratogenicity based on the Rothman et al. (1995) paper. The EU did not use any uncertainty factor because it believed that data from other studies supported a true threshold of more than 3,000 μg/day and that this number covered the risk of hepatotoxicity.

Using the same paper, the IOM study committee determined the no-observed-adverse-effect level (NOAEL) for teratogenicity to be 4,500 μg/day and used an uncertainty factor of 1.5 to establish a UL of 3,000 μg/day. However, the IOM study committee had already decided to use 14,000 μg/day as the LOAEL for liver toxicity, with a high uncertainty factor of 5 because of the severity of the effect, resulting in a UL of 3,000 μg/day. Because the study committee believed it would be confusing to have women of reproductive age with one UL and all others with another UL, it somewhat adjusted the numbers to come out with the same UL.

by phytate intake, a confounding factor. For the Estimated Energy Requirement (EER), the goal was to maintain a healthy weight at an acceptable level of physical activity. That is, the EER was based on energy balance (no weight gain), not on reduction of disease risk—a different type of paradigm than originally envisioned.

Selection of Endpoints

In general, the selection of endpoints was based on data availability. For ULs, the endpoints were frequently concerned with public health protection, often using a benign adverse effect (e.g., flushing rather than liver dysfunction) to be more protective. The selection of endpoints was not based, for the most part, on the strength or consistency of evidence or on the severity or clinical importance of the endpoints. When the (ideal) data were lacking, the study committees still had to provide numbers; it was emphasized that "no decision was not an option." This is because the numbers are needed for so many purposes, such as goals for individuals, dietary assessment and planning, food fortification, food assistance program evaluation, food labeling, agricultural policies, dietary guidance policies, and educational program planning.

In the future, it might be better to select endpoints more scientifically. The use of biomarkers that correlate with a disease or physiological state would be very helpful. The biomarkers should be attributable and responsive to the nutrient in question—key questions that can be answered using SEBRs. Further SEBRs allow the ranking of the quality of the evidence according to the degree of confidence in the conclusion. If the biomarker is found to be valid, the dietary intake can be correlated with the biomarker, and the overall quality of the data can be ranked.

Systematic Evidence-Based Reviews

SEBRs can answer only limited types of questions.[4] Nevertheless, they are independent and unbiased reviews of a defined topic by a group with no stake in the outcome. They can account for confounders (e.g., dietary supplements) in ranking. They can determine the validity of extrapolations or interpolations. They can increase the transparency of decisions made about specific endpoints, which increases the replicability of the data by other groups. The importance of the SEBR is illustrated in Box 2-3.

Other Challenges

One quandary for application is that sodium, potassium, calcium, vitamin D, vitamin E, and linoleic acid DRIs are unrealistic values, given the North American food supply and dietary habits. Almost no one meets the numbers for these nutrients. While the science for setting DRI values takes precedence and should not be compromised because of real or perceived inconsistencies about what the population is eating, DRI reports may need to include more discussion about these problems when they occur.

While decisions about the use of DRIs for nutrition labeling are outside the purview of the DRI development process, related issues raise interesting questions, such as what to do if there is no DRI (e.g., *trans* fat), what to

---

[4]SEBRs are discussed in further detail in a separate presentation later in this chapter.

BOX 2-3
β-Carotene Case Study and the Evidence-Based Review

The β-carotene trials were started on the basis of many epidemiological studies showing that the higher the β-carotene in the serum or diet, the lower the incidence of lung cancer in smokers. However, when an intervention trial was done with β-carotene at a fairly high dose, more lung cancers, not fewer, were found in the β-carotene group (Heinonen and Albanes, 1994). This was backed up by a second trial in the United States, the CARET trial, done in 1996 (Omenn et al., 1996).

Three years before the first of these trials, in 1991, the Food and Drug Administration (FDA) had looked at the large number of available studies (mostly retrospective or prospective epidemiological studies) with either cancer or premalignancy as the endpoint. The first criterion used to evaluate the studies was: Did they allow attribution of β-carotene per se to the observed health effects, not simply to diets or dietary patterns that were rich sources of these nutrients or to serum/plasma levels that could be markers of diets rich in these nutrients? The second criterion was: Did they provide a sufficient basis for relating intakes to the actual reduced risk of cancer (because there were no validated biomarkers at the time to serve as surrogates for cancer sites)?

The bottom line was that the FDA's systematic evidence-based review (SEBR) led it to reject the health claim that antioxidants collectively and carotene specifically could protect against cancer. The government might have saved itself considerable expense if it had paid attention to the FDA's SEBR performed 3 years before the huge intervention trial began.

do if there is an AI (e.g., calcium), how to identify a single dietary value if there is a distribution range, and how to choose between an EAR and an RDA. It should be remembered that people use food labels to choose among food products, not to formulate their diets.

Whether an approximate (e.g., interpolated) EAR that is scientifically based can be derived when the data are nonexistent or inadequate should be investigated. If it can be derived, the best way to express that value to make it more useful should be determined. Consistent guidelines should be developed for setting uncertainty factors and for rating the overall evidence for a DRI value, based on the strength of the data, the consistency, the public health relevance, and the applicability to the person or persons of interest.

Usefulness of the DRI Framework and Conclusions

The DRI framework has often been found not to be useful for planning for groups, such as WIC, primarily because too many assumptions have to be made (e.g., the distribution of intakes will not change with a particular intervention). For planning for individuals, it is questionable how the RDA is to be used. The RDA is probably most useful as a goal that is either met or not met. For assessing individual dietary adequacy, the probability equations have been found to be too cumbersome to use; as a result, only 5 percent of dietitians admit to using them. However, for assessing intakes of groups, such as WIC, the framework has worked well. In summary, the pros and cons of the past paradigm are listed below.

Pros

- A comprehensive review of scientific literature at the time was performed.
- A risk assessment model was developed.
- The framework for assessing group dietary intakes worked well using the EAR cutpoint method for prevalence of inadequacy.

Cons

- For the most part, the health endpoint data on which to base DRIs were lacking.
- Variance data were lacking.
- It was necessary to make many extrapolations, the scientific validity of which was unknown.
- Long-term data were limited.
- The uncertainty factors for deriving ULs were very subjective.

CASE STUDY: APPLYING THE DRI FRAMEWORK TO CHRONIC DISEASE ENDPOINTS

Presenter: Paula Trumbo

The conclusion that the "*reduction in risk of chronic disease is a concept that should be included in the formulation of future RDAs where sufficient data for efficacy and safety exist*" (IOM, 1994) had a notable impact on the DRI development process. It influenced the way in which nutrients were grouped for review, as noted in the following examples:

- Calcium and related nutrients were grouped together because of their role in bone health and general health.
- Antioxidants were reviewed together because of their potential role in reduction of risk of chronic diseases, such as cancer and CHD.

- Electrolytes were grouped because of their role in blood pressure and hypertension.

Moreover, a guiding principle that was conveyed to the DRI study committees was the need to review the evidence on chronic disease first to determine if it was possible to use such data to set a DRI.

Setting EARs Based on Chronic Disease Endpoints

Of the nutrients that were assigned reference values related to nutritional adequacy, only five were based on chronic disease endpoints. While the DRI study committees were encouraged to set an EAR rather than an AI because of the limited utility of the AI for assessment purposes, the reference values related to nutritional adequacy that were developed for nutrients based on chronic disease endpoints were all AIs. The endpoints were

- osteoporosis and fractures for calcium and vitamin D, as well as balance data and biomarkers for vitamin D;
- dental caries for fluoride;
- CHD for fiber; and
- a combination of endpoints, including salt sensitivity (a risk factor of hypertension), kidney stones, and blood pressure, for potassium.

An important question to ask is "Could EARs have been set using chronic disease endpoints if sufficient data had been available?" The EAR is an average daily nutrient intake level that is estimated to meet the requirement (defined by the nutrient-specific indicator or criterion of adequacy) of half the healthy individuals in a subpopulation. In Figure 2-1, at a very low intake of 30 units for nutrient X, there is a risk of inadequacy in 100 percent of the subpopulation. At an intake level equivalent to the EAR of 100 units, the risk of inadequacy is 50 percent. At an intake level of approximately 140 units, there is only a 2–3 percent risk of inadequacy for nutrient X (i.e., the RDA).

This DRI paradigm worked well when the EAR was based on essentiality because nutrient-specific indicators were being used, such as balance data for molybdenum, factorial data for iron and zinc, status biomarkers that were unique to copper and vitamin E, and turnover data for iodine and carbohydrate. Furthermore, endpoints of inadequacy could be used to set an EAR because all individuals are at risk of inadequacy for essential nutrients.

The challenge in fitting a chronic disease endpoint into this DRI paradigm is illustrated by a clinical trial that evaluated potassium intake and frequency of salt sensitivity. This trial provided multiple doses of potassium

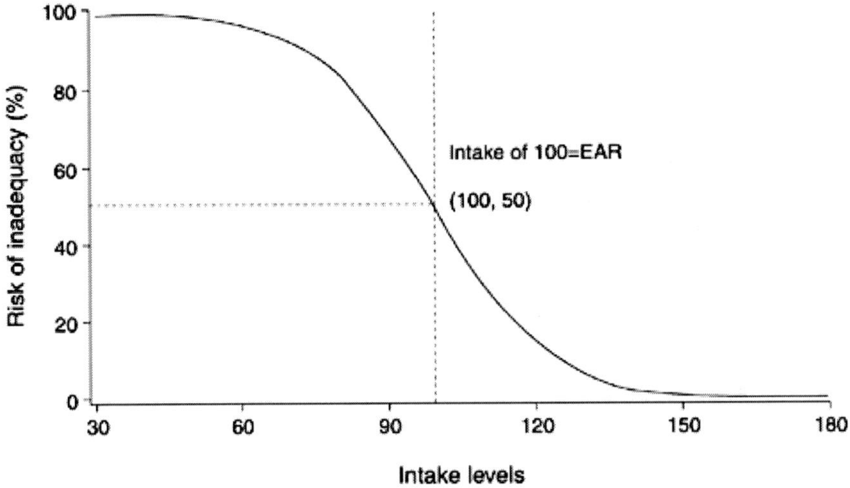

FIGURE 2-1 Estimated Average Requirement for hypothetical nutrient X.
NOTE: EAR = Estimated Average Requirement.

to individuals who were consuming very high levels of salt. The highest frequency of salt sensitivity occurred at a very low level of potassium intake (30 mmol/day) and was about 78 percent for African Americans and 37 percent for Caucasians (Figure 2-2). It became obvious that it was difficult to apply data from such a trial to the DRI paradigm, which assumed that the risk of inadequacy at very low intake is 100 percent for the population.

If the EAR is to be based on chronic disease risk reduction rather than reduction of the risk of nutrient inadequacy, then the definition of the EAR would be the nutrient intake level to reduce the risk of chronic disease in half the healthy individuals in a particular subpopulation, or to achieve an absolute risk reduction of 50 percent (where absolute risk is the probability of getting a disease over a certain time and is affected by the relative risk of a particular risk factor, such as intake of an individual nutrient).

Each component in absolute risk reduction has challenges. One is the assumption that the absolute risk of a chronic disease is 100 percent for a subpopulation, as is the case for risk of inadequacy based on essentiality. Perhaps this is the case for dental caries, but it is not the case for other disease endpoints, such as osteoporosis, CHD, and kidney stones. The absolute risk of osteoporosis is not 100 percent, even for Caucasian postmenopausal women, and the absolute risk for CHD is even less than that for osteoporosis. The prevalence of kidney stones is approximately

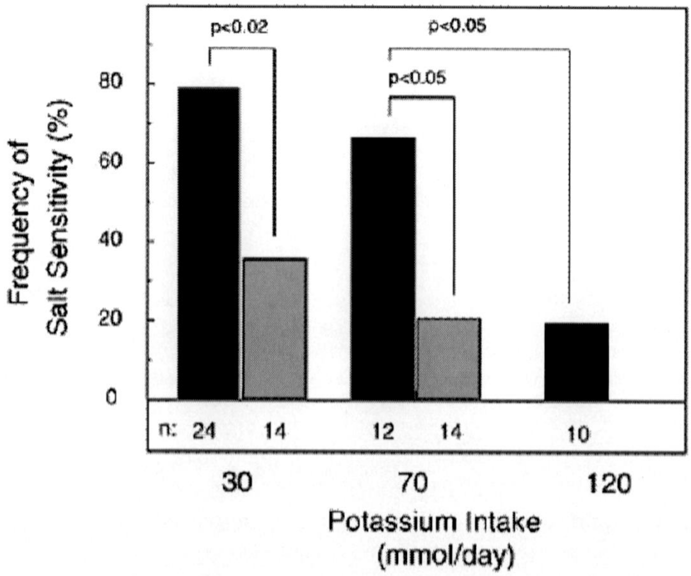

FIGURE 2-2 Effect of potassium intake on frequency of salt sensitivity in nonhypertensive African American men (solid bar) and white men (gray bar).
SOURCE: Morris et al. (1999). Normotensive salt sensitivity: Effects of race and dietary potassium. *Hypertension* 33(1):18–23.

5 percent, and the prevalence is even less for certain individual cancers, depending on the type.

The other challenge is observing a chronic disease risk reduction of as much as 50 percent in response to the intake of an individual nutrient. Chronic diseases are not nutrient specific. Rather, they are multifactorial, with other factors, such as genetics, age, environment, lifestyle, and other nutrients, contributing to the risk. Unlike the effectiveness of reducing the risk of a nutrient deficiency, risk reduction of most chronic diseases by diet is limited.

For instance, although one of the endpoints considered for calcium was fracture risk, the DRI study committee chose to reject the observational data on fracture risk because of the influence of confounding factors. One reason given for not setting an EAR for vitamin D was that it could not account for the contribution of sunlight exposure, which is affected by a wide variety of factors (this would also influence reference values related to nutritional adequacy based on essentiality). For dental caries, while the absolute risk is probably at or near 100 percent in North America, the DRI study committee on macronutrients stated that caries occurrence was influenced by frequency of meals and snacks, sugar products, content of foods, oral hygiene, and exposure to fluoride.

In general, individual nutrients do not usually yield a relative risk reduction as high as 50 percent. The AI for fiber was based on three prospective cohorts, with a relative risk reduction for CHD ranging from 16 to 41 percent. The greatest risk reduction was only 41 percent, because CHD, like other chronic diseases, is multifactorial. An individual nutrient would not be expected to result in a risk reduction of that magnitude. The AI for potassium was based, in part, on the risk of kidney stones, where three prospective cohorts yielded a relative risk reduction ranging from 21 to 51 percent. Thus, while one of the three prospective cohorts observed a 51 percent risk reduction at the highest quintile of potassium intake, the weighted average relative risk reduction would fall short of 50 percent.

Another complication was the macronutrients, particularly fat and carbohydrate, because of their interrelatedness in the diet. Although the DRI macronutrient study committee tried to define specific reference values related to nutritional adequacy for the individual macronutrients, it became obvious this was not possible. Thus, the AMDRs were developed and set for the macronutrients, and some of them were based, in part, on risk biomarkers of chronic disease, such as CHD for fat. In summary, the challenges of using chronic disease endpoints for setting an EAR/RDA are

- a nutrient-specific indicator is not being applied;
- the absolute risk of most chronic diseases applies to only a portion of the population; and
- achieving risk reductions as high as 50 percent is very difficult for most chronic diseases because of the multifactorial nature of chronic diseases.

Therefore, the definition of an EAR does not allow for the use of chronic disease risk reduction in setting recommended intake levels, which is an opinion shared by many who have worked closely with the DRI process.

Setting ULs Based on Chronic Disease Endpoints

Chronic disease endpoints have also been used to set ULs. A UL could be set if sufficient data were available for identifying a LOAEL or, preferably, a NOAEL. For essential nutrients, only one UL was set based on a chronic disease endpoint: sodium and blood pressure (a surrogate endpoint for cardiovascular disease [CVD]). A NOAEL could not be identified because of the lack of a threshold, and it was not known if blood pressure would continue to drop below the lowest sodium intake level provided (50 mmol/day). However, because sodium is essential, an AI was set based on factorial data (65 mmol/day). The UL was set based on a LOAEL of 100 mmol/day, even though a threshold was lacking.

For nonessential macronutrients, such as *trans* fat, cholesterol, and saturated fat, there was also no observed threshold effect using risk biomarkers of CHD. As intake of any of the three macronutrients decreased, the biomarkers of heart disease (e.g., change in total cholesterol, change in LDL/HDL cholesterol) continued to decrease. The lowest intake levels approached zero for percentage change in these risk biomarkers. Because these three macronutrients are not essential, the UL should be 0 percent of energy; however, this level would have required extraordinary and unachievable changes in dietary patterns. Therefore, a UL could not be set for these three macronutrients.

Implications for Reference Values Related to Nutritional Adequacy

The challenges of setting reference values related to nutritional adequacy based on chronic disease risk reduction were recognized when the framework for revising the RDAs was being considered. A 1994 IOM report stated that "*If reduction of risk of chronic disease is to become a criterion in the development of future RDAs, many questions must be faced*" (IOM, 1994). Some of these questions and associated comments follow:

- "How can concerns regarding potential interactions among nutrients be addressed?" This could include the interaction of nutrients that are confounders of disease risk.
- "Should levels of nutrient intake be expressed in terms of numerical ranges, in terms of food patterns, or in some other way?" Numerical ranges (AMDRs), rather than a specific intake level, were set for the macronutrients.
- "How can desirable levels of intake be extrapolated for groups not included in clinical trials (such as children, adolescents, young adults, and the elderly)?" Gender and, most often, age can be confounders of disease risk.

Furthermore, at the 1993 FNB meeting that preceded the 1994 IOM report, some commenters argued that "the RDAs should remain distinct from the dietary guidelines for reducing the risk of chronic disease."

Despite the limitations in the use of the AI, setting AIs based on chronic disease worked rather well. This is because a prescriptive approach was not being used to derive AIs as it was for setting EARs and therefore RDAs. Another issue is that AIs can be based on observed or experimentally determined estimates of intake (i.e., observational studies that alone were sufficient for setting AIs, but not EARs). Furthermore, the AI is expected to meet or exceed the amount needed to maintain a defined nutritional state or criterion of adequacy for essentially all members of a specific subpopulation (i.e., the RDA). Along with that, the reference values related to nutritional adequacy based on chronic diseases would be expected to be greater than reference values related to nutritional adequacy based on the daily requirement for many essential nutrients (e.g., if an RDA for potassium had been set based on essentiality, it would have been much lower than the AI of 120 mmol/day based on chronic disease risk reduction).

Possible approaches for addressing chronic disease endpoints in terms of reference values related to nutritional adequacy include the following:

- Continue to set AIs based on clinical/observational data.
- For the macronutrients, continue to set AMDRs, particularly in the lower range, based on clinical/observational data, as well as dietary intake data.
- Develop a new criterion/DRI that provides a prescriptive way to set recommendations based on chronic disease endpoints.

In addition, the approach used to set the upper range of the AMDRs might be useful in setting a maximum intake level for nonessential nutrients without a threshold or NOAEL by relying on clinical/observational data and dietary data (e.g., menu modeling and survey data).

CASE STUDY: APPLYING THE DRI FRAMEWORK TO NON-CHRONIC DISEASE ENDPOINTS

Presenter: Allison Yates

Discussions about the experience of using non-chronic disease endpoints (specifically adequacy status endpoints) to establish DRIs benefit from acknowledging some underlying realities. First, the reality is that "no decision is not an option," meaning that the absence of some type of DRI value leaves a scientific gap and is problematic for users, particularly government policy makers. The DRI process therefore focused on deriving a value whenever possible or offering good justification when it was not possible. Second, endpoints reflect a "continuum of adequacy" for every nutrient, whether to prevent a frank deficiency state or a chronic disease. The expectation is, particularly in deriving reference values related to nutritional adequacy, that a quantitative determination of adequacy can be developed based on a validated biomarker methodology with a dataset (the evidence-based component).

Key components of the DRI framework are as follows:

- A criterion of adequacy or excess based on decreasing the risk or a validated biomarker with strong evidence

- An EAR based on a reliable dose–response, so that half of the individuals have inadequate intakes
- Primary endpoint data from more than one laboratory
- A UL based on chronic intake and a serious adverse effect

To describe the experience of using non-chronic disease endpoints, three nutrients are highlighted in this discussion:

- Vitamin C, an example of an antioxidant with a known continuum of adequacy
- Iodine, as an example of a deficiency state that has significant public health significance in many parts of the world today
- Vitamin K, as an example of a nutrient with poorly characterized intake and requirements when compared with other nutrients

Antioxidants: Vitamin C

The study committee on dietary antioxidants and related compounds not only was asked to develop dietary reference levels of intake, but also was given other tasks. They were defining dietary antioxidants, reviewing the scientific literature on the antioxidants and selected food components that may influence their bioavailability, addressing the safety of high intakes, and providing guidance on uses of the developed reference intakes.

The study committee defined dietary antioxidant as a substance in foods that significantly decreases the adverse effects of reactive species, such as reactive oxygen and nitrogen species, on normal physiological function in humans. In addition to vitamin C, vitamin E, and selenium the study committee examined data about β-carotene and other carotenoids (α-carotene, β-cryptoxanthin, lutein, lycopene, and zeaxanthin).

The continuum for vitamin C is shown in Figure 2-3. Very low levels of vitamin C are required to prevent scurvy. Bleeding gums occur at a slightly higher level. Urinary excretion is observed at about 60 mg in urine. Although many have evaluated the effect of vitamin C on chronic disease, that endpoint was not chosen by the panel. Diseases associated with increased levels of reactive oxygen and nitrogen species include age-related eye disease, atherosclerosis, cancer, CHD, diabetes, inflammatory bowel disease, neurodegenerative disease, respiratory diseases, and rheumatoid arthritis. All of these have other causative factors in the diet and the environment, and genetics plays a major role, which makes it difficult to use these as criteria of adequacy.

Possible biomarkers for vitamin C include inhibition of superoxide in neutrophils, oxidative deoxyribonucleic acid (DNA) and chromosome damage, immune markers, and relationship to chronic disease outcomes.

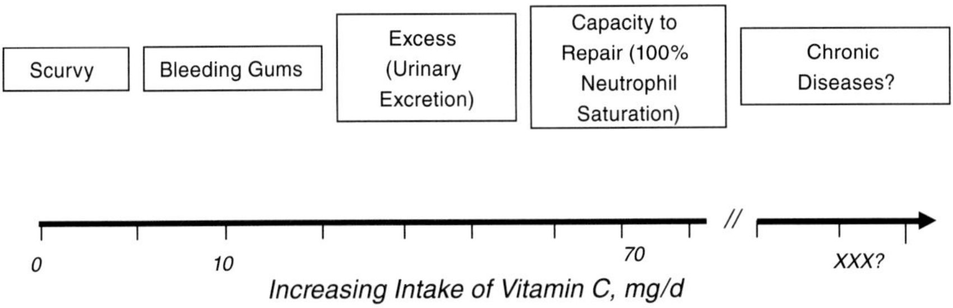

FIGURE 2-3 Vitamin C endpoints.

What was chosen as the indicator of adequacy for adults was the capacity to repair (neutrophil saturation with ascorbate and the ability to deal with superoxide compounds) at a level of about 70 percent saturation, which was also the point at which urinary excretion of vitamin C became appreciable.

The study committee wanted scientifically valid experiments. It was looking to measure relevant biomarkers that were significantly related to the disease endpoints, were based on in vivo experiments, and played a role in health. It also wanted reliable intake data. What it did not want were strictly observational data, strictly antioxidant-type functions, or overreliance on animal data and associations, rather than causation.

The findings can be found in the IOM report on vitamin C and other related nutrients (IOM, 2000b). The rationale for the recommendation for vitamin C was that there was no accepted methodology comparing vitamin C intake with an in vivo antioxidant effect; they could find vitamin C functioning as an antioxidant in white blood cells or neutrophils, but there were no data relating that to intake; and there were data relating leukocyte ascorbate levels to liver and body pools of ascorbate.

EARs and RDAs were developed for children and adults as well as pregnant and lactating women. Most of the values were based on extrapolation from data from one study in men with a small sample size. Research recommendations were made, indicating that more data are needed in certain areas, including the establishment of a reliable functional biomarker, interaction of vitamin C and iron, and the effect of vitamin C supplements on the fetus.

The IOM (2000b) applied the EAR cutpoint methodology to vitamin C intake from the National Health and Nutrition Examination Survey (NHANES), showing that the intakes of 10 percent of women and 21 percent of men were below the EAR. The value of the EAR is that one could assume there was a similar percentage of lower levels of saturation of ascorbate and ability to deal with superoxides, not that scurvy itself was of concern.

Micronutrients: Iodine

Iodine is an essential component of the thyroid hormones involved in the regulation of various enzymes and metabolic processes. The continuum for iodine (Figure 2-4) goes from cretinism at very low levels through iodine accumulation to higher levels of urinary excretion.

The iodine EAR for adults is based on iodine accumulation (IOM, 2001). Three studies were available on thyroid iodine (radioiodine) accumulation and turnover in adults, but they were limited by small sample sizes. The requirements were 96.5 µg/day (n = 18) (Fisher and Oddie, 1969a), 91.2 µg/day (n = 274) (Fisher and Oddie, 1969b), and an absolute iodine uptake of 21 to 97 µg/day (n = 3) (DeGroot, 1966). The study committee on micronutrients selected turnover as the basis for the requirement and calculated an EAR of 95 µg/day, which was assumed to be adequate for about half the individuals.

This EAR for adults 19–50 years of age was extrapolated to other parts of the population (e.g., >51 years). For children 1–3 years, an iodine balance study on nutritionally rehabilitated children 1.5–2.5 years of age (Ingenbleek and Malvaux, 1974) gave an EAR of 65 µg/day. As the EAR extrapolated from adults would be ~36 µg/day, the study committee used the balance study as a basis for the EAR, as it resulted in a higher estimate. The same occurred with the age group 4–8 years, but the EAR was based on a different iodine balance study (Malvaux et al., 1969). In the case of children 9–13 years of age, the actual iodine balance data in children that age (Malvaux et al., 1969) resulted in an EAR of 55 µg/day. As extrapolation from adults gave an EAR of 73 µg/day, the study committee used the more protective higher estimate.

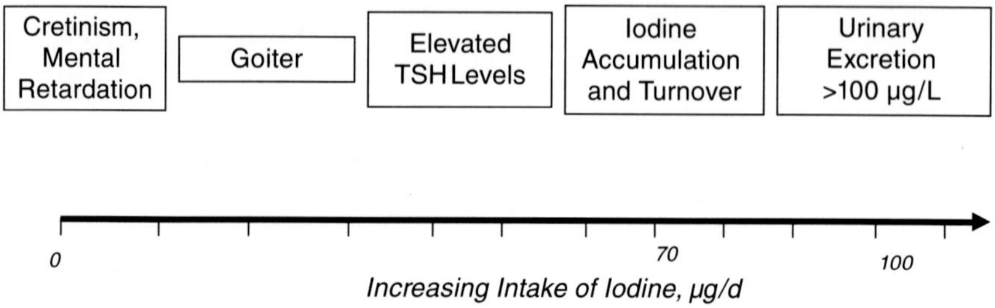

FIGURE 2-4 Iodine endpoints.
NOTE: TSH = Thyroid stimulating hormone.

Vitamin K

Not much information is available on vitamin K, which is required as a coenzyme for the synthesis of proteins active in blood coagulation and bone metabolism. Dietary intake data were available to the micronutrient study committee, but the order or continuum of endpoints considered (including possible relationships with osteoporosis and atherosclerosis) had not been identified (Figure 2-5). The median intake from NHANES III was the basis of the AIs for children 1 year and older to adults.

Conclusion

The major challenges experienced in setting reference values using non-chronic disease endpoints result from the existence of a continuum of adequate levels of intake reflective of the possible endpoints that could be selected. This continuum is different for different nutrients. Moreover, the quantitative determination is critical, regardless of what nutrient is being considered. Finally, challenges always arise when scientific judgment must be used; it is used frequently when data are limited, yet there is the clear need to derive reference values so that policy decisions can be based on some scientific data as opposed to no scientific data. DRI development is a long-term, iterative process, and so we should expect that new data will provide new answers.

DISCUSSION: FRAMEWORK PROS/CONS; CASE STUDIES

Co-Discussants: Patsy Brannon and Alice H. Lichtenstein

The session moderator, Dr. Stephanie Atkinson, introduced the discussants and invited each one to offer an opening remark.

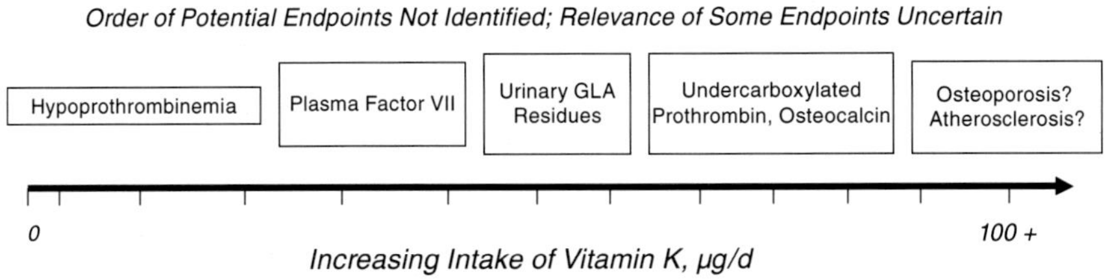

FIGURE 2-5 Possible vitamin K endpoints.
NOTE: GLA= γ-Carboxyglutamic acid.

Discussant Opening Remarks

Dr. Brannon opened the discussion by reflecting that our concept of good nutrition has evolved as nutritional and biomedical sciences have advanced. Because we now think in terms of decreasing risk of disease as well as eliminating deficiencies, the DRI framework has become much more complex, and it may become even more so as our understanding of the interactions among genetics, nutrients, diet, and environment continues to increase. She raised several important issues: the need to make decisions based on limited data; the need to determine whether the data have evolved sufficiently to allow EARs to be estimated for the nutrients for which AIs were established; the value of conducting SEBRs; the setting of priorities between innovative research and research that deepens our understanding but does not necessarily advance our knowledge; and the use of the risk assessment model.

Dr. Lichtenstein mentioned additional issues concerning the need to (1) address requirements for single or very similar nutrient groups rather than for such large groupings of very different nutrients; (2) foster study committees' ability to consider the unique aspects of each nutrient; (3) reconsider life stage groupings for nutrients, given that the values are needed even in the face of limited data and more data are available for some life stage groupings than for others; (4) increase understanding about the nature of the goals the DRIs reflect, particularly that some values reflect a goal to increase intake (vitamin A) and others a goal to decrease intake (cholesterol); and (5) encourage further discussion about the multiple causative factors for chronic diseases, particularly as it may relate to separating dietary patterns and individual nutrients.

General Discussion

Drs. Russell, Trumbo, and Yates joined the discussants at the dais, and a brief group discussion took place. The initial focus included macronutrient recommendations, the target population for DRIs, the ability to achieve DRI intake levels through typical diets, and the value in constituting a single study committee to develop both EARs/RDAs and ULs.

Macronutrients

One participant noted that quantitative reference values were not given in the reports on macronutrients; rather, advice such as "intake should be as low as possible while consuming a nutritionally adequate diet" was provided. She suggested that a quantitative value would have been useful and could have been developed through the use of modeling techniques. Another participant suggested that a different approach for developing

DRIs for macronutrients may be needed as compared with that for vitamins and minerals. Furthermore, efforts to improve communication about macronutrient recommendations should be pursued. For example, a recommendation for "up to 25 percent of energy" for sugar intake could be misinterpreted as the target goal for intake.

Target Population

The moderator noted that the morning's presentations included the issue of setting DRIs for "healthy populations" and asked whether obesity would be viewed as a disease. Another participant commented that overall there was a need to establish a "representative state of health" that might serve as a starting point for DRI development. Nonetheless, there was agreement that this issue represents a quandary.

DRI Recommendations Versus Estimated Intake

A participant asked about the appropriate strategy when a scientifically valid reference value for intake cannot be achieved realistically with the current diet. One response was that the answer would depend on whether the endpoint selected for the value reflected a public health concern or not, and that consideration should be given to the level of risk to be tolerated, with the view that the EAR is the median estimated requirement. The discussion continued, focusing on supplementation as a solution, noting specifically the possible role of *targeted* supplementation. These comments led one participant to remark on the value of tasking a single study committee with responsibilities for both EAR/RDA and UL development because these reference values at some point become highly related. It was noted that during DRI development the groups had collaborated; however, given the amount of information to review, those responsible for EARs or ULs were often unwilling or unable to review other draft sections of the report.

Dietary Patterns

When the discussion was opened to all members of the audience, one participant noted that during the past 10 years, the development approach moved from specifying a "black-and-white" cutoff in the form of an RDA to consideration of a probability model. This approach made it clear that there was a distribution of requirements in the population. Given this probability paradigm and the interest in dealing with chronic disease, he suggested that consideration of dietary patterns would be useful and could address in part the multiple causes of disease; it also would have more direct clinical and practical application. Furthermore, advances in genetics and nutrigenomics may follow the identification of groups of people who are at greater risk or who will benefit from a modified pattern of nutrient intake. Another audience member pointed out that confounding may be introduced by wide use of dietary supplements.

Endpoint Continuum

One participant remarked that there is no clear distinction between a chronic disease endpoint and a more traditional adequacy endpoint. Instead, all endpoints are part of a continuum. He also noted the need to focus on the way changes in nutrient intake impact biomarkers as well as nutrient interactions. Another participant preferred the use of physiological endpoints rather than chronic disease endpoints. He offered the example of dark adaptation. It is not a disease but a dysfunction of physiology, and such a condition can continue for years without progressing to the next stage.

Estimating Intake

With regard to the activities for setting DRIs, one commenter indicated that it is important to consider the effects of measurement error (imprecise measurement of the true usual intake) in self-reports of dietary intake, particularly when food frequency questionnaires are used. In general, failing to adjust for measurement error causes problems in estimating relationships between diet and health outcomes by attenuating the true relationship between diet and the outcome.

AIs and Chronic Disease

A participant admitted to being a vocal opponent of AIs, but was persuaded to consider using AIs when chronic disease endpoints are used, as other countries have done. She questioned whether it would be possible to have EARs and RDAs for essentiality and perhaps an AI for the same nutrient as appropriate for reducing the risk of chronic disease. It was considered possible, but there would be ramifications for user applications, including considerable need for information and education.

TWO PERSPECTIVES: THE DRI FRAMEWORK

Perspective I

Presenter: George Beaton

I have watched the evolution of DRIs and their application, with the sense that I have been responsible for part of it. I now stand before you to say that we have gone too far. The probability approach that everyone espouses, and which I promoted, continues to be a useful concept. However, in terms of its current application, particularly to individuals, we have gone far beyond the data. We must retreat a bit and ensure that we ground ourselves in science.

There will be negative remarks at this workshop, of course—what the DRIs did *not* accomplish. Yet it is important to recognize what the DRIs *did* accomplish. First, they are probably the most comprehensive literature review in nutritional science to date. Second, they provide formal recognition of the importance of the EAR and the fact that we cannot escape dealing with distributions. Failure to recognize this reality is a major cause of confusion and controversy in the nutrition community. Third, the introduction of the UL was a major breakthrough because our community tends to believe that more is always better and thus encourages the use of often unnecessary supplements. We must build on what we have accomplished in the DRI process and move forward.

The sole purpose of the DRI development process is to foster the *application* of nutritional science. If this activity advances that science or promotes research, so much the better. But that is not its purpose. Rather, the goal is to apply the principles of scientific analysis throughout the process. There will always be issues of judgment, not to mention individuals with strong views. But the most important task is to ground the DRI process in the principles and concepts of the scientific method.

From the views being expressed by users, it is clear that everyone wants a single number—not distributions or ranges—that fits their application. However, they have to remember that the numbers are not the same for all the applications. It has been a problem from the beginning that the number needed *should* differ among applications.

The original driving force behind reference values or nutritional standards was to try to plan food supplies for war-torn countries and then survivors of prison camps. This purpose was carried on by the Food and Agriculture Organization of the United Nations and by agriculture ministries in, for example, the United States and Canada. Many other applications now exist, including food programs, nutrition labeling, and individual counseling. Each application is different. It is *impossible* to develop a single number that would fit all applications. It *is* possible to develop core parameters of the requirement and adverse effect distributions that will allow the development of evidence-based approaches to the diverse applications.

The DRI process should not attempt to provide derived reference values for all applications. Only the core values that are absolutely needed should be derived. These are the central tendency of the requirement distribution, the EAR, and the tolerable upper level, or UL. Groups with relevant expertise could then be convened to provide guidance on how to use the core values for specific types of application. The IOM committee that dealt with nutritional labeling and fortification (IOM, 2003b) is an excellent example of providing guidance on the application of DRIs in particular circumstances. That should be seen as the model of the future.

For the DRIs of the future, we should consider what core estimates must be available to serve the diverse needs of users. Having established these core needs, meeting them should be the primary objective of the exercise.

Equally important, we *must* consider ways in which we can evaluate and validate DRI requirement estimates. We need biomarkers of requirement that are meaningful and measurable in the field (e.g., refer to the attempted validation of the iron requirement estimates in the DRI report on iron [IOM, 2001]).

The desirability and ultimate utility of examining the "reality" of the application of the DRI values can be illustrated with protein. Based on NHANES III protein intake data as used in the DRI reports to describe distributions of usual intake, I conducted dietary assessments using the methods provided as guidance for users. When the prevalence of apparently inadequate protein intakes as grams per kilogram body weight per day were considered by age, gender, and usable protein (taking into account likely digestibility and amino acid score, both of which fall as vegetable source protein increases in the diet [IOM, 2002/2005]), the results were surprising. No problem was apparent for youngsters, who are supposed to be vulnerable, but there was an approximately 25 percent prevalence of inadequate protein intakes in the older adult groups (over 50 years of age).

Since the report on macronutrients suggested a major increase in lysine requirement which would affect the amino acid score, further examination was undertaken. Individuals were classified by vegetable protein intake constituting less than or more than 50 percent of the total dietary protein. For each subgroup, the estimated prevalence of protein inadequacy was estimated (Figure 2-6). A shocking 63 percent of women over age 51 appeared to have inadequate intakes if they consumed more than 50 percent of intake from vegetable sources (these persons were assumed to be vegetarians). This might imply a major nutritional problem among an identifiable subgroup of the United States population, a problem warranting a high priority for action.

However, there is a serious quandary. These analyses were based on requirement expressed as grams per kilogram body weight per day, the original unit in which protein requirements were estimated. The DRI study committee requested that requirements be presented as grams per day applied to reference individuals (omitting any provision for variation in body size). I compared these modes of expressing protein requirements, specifically (a) grams per kilograms per day, (b) the grams per kilograms per day referred to relevant reference persons, a man weighing 70 kg and

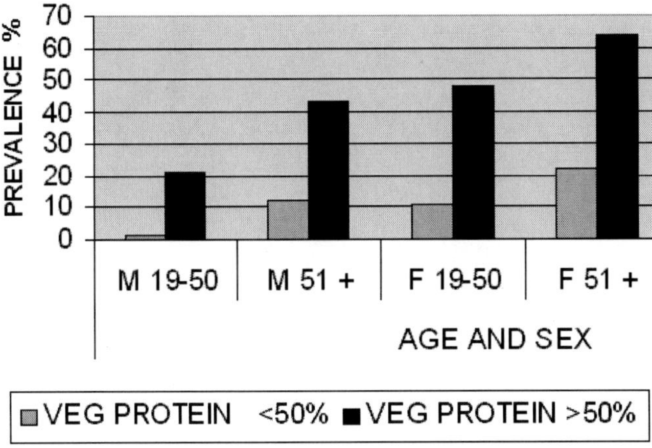

FIGURE 2-6 Impact of choice of dietary protein source on protein inadequacy. NOTE: High vegetable protein group = 12 percent of subjects.

a woman weighing 57 kg, and (c) the suggested lower limit of the AMDR (IOM, 2002/2005). The apparent "problem" of inadequate intakes was seen only when requirement was expressed as grams per kilograms per day (Figure 2-7). For females 19–50 years of age, the prevalence of apparent inadequacy dropped from 14 percent to 3.3 percent, and even lower using the AMDR lower limit (10 percent of energy as protein). Equally dramatic effects would be seen in the apparent prevalence of inadequacy among women over 51 years consuming vegetarian diets shown in Figure 2-6.

We are left with three different estimates of the apparent adequacy of protein intakes among adults in the United States. These range from the inference of the existence of a major public health problem in an identifiable subgroup of the population, to satisfaction that protein intakes are adequate for nearly all persons. *But which estimate is valid?* How do we determine the "truth" using independent measures? We have no field-applicable measure equivalent to the nitrogen balance criterion used to estimate requirements. *That is a most unsatisfactory situation.* Unfortunately, parallel situations hold for several other nutrients. How do we validate the estimated prevalence of inadequate intakes if we cannot measure prevalence by direct assessment? These issues are important both for national and regional planning and for scientific validation of requirement estimates, but we have not developed the concepts and tools we need to address them. This is an urgent need for future DRIs.

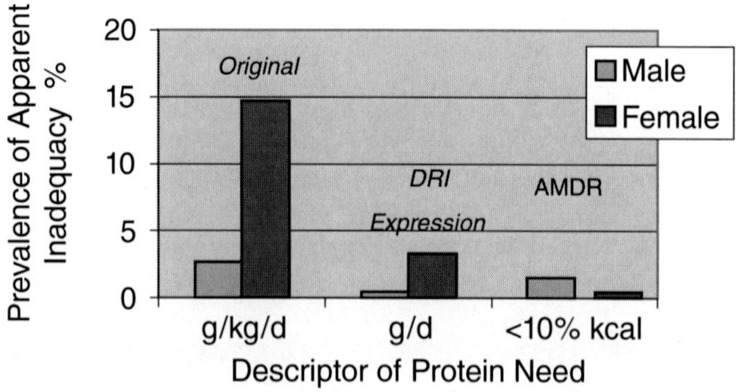

FIGURE 2-7 Impact of choice of assessment criterion for protein, adults 19–50 years of age.
NOTE: DRI = Dietary Reference Intake; AMDR = Acceptable Macronutrient Distribution Range.

From my perspective, there are dreams about the things we would like to have for use in developing future DRIs, shown on the left in Table 2-1, and there are the realities of whether we can generate them with a foundation in science, shown on the right of the table.

Much depends on the precision wanted/needed in any final application. It also depends on how far we are prepared to abandon science in favor of opinion and judgment.

In the end, we are constrained by a realization that what we dream is

TABLE 2-1 Dreams Versus Realities of the DRI Process

Desired Information	Feasibility of Obtaining This Information
Requirement distribution midpoint (EAR[a])	Yes (for some)
Full requirement distribution (e.g., CV[b])	No (for all except at very high cost)
Midpoint intake of detrimental effect distribution	No (for all)
Judged start of risk of detrimental effect (UL[c])	Yes (for some)
Usual intake distribution, groups and populations	Yes (if data collected)
Usual intake, individual	No (for all)
Correlation between intake and requirement	No (for all)

[a]EAR = Estimated Average Requirement.
[b]CV = coefficient of variation.
[c]UL = tolerable upper intake level.

desirable and helps drive scientific advancement, but what we must deal with is reality. We must temper our theories, approaches, desires, and dreams with reality. We must always remember that the whole purpose of any DRI process must be to come up with evidence-based information that can be applied to real life.

Perspective II

Presenter: Janet King

My perspectives on the DRIs come from two experiences: as chair of the FNB in 1994 and as chair of the Dietary Guidelines Advisory Committee in 2005.

In 1994, we began the process of revising the RDAs, defined as "the levels of intake of essential nutrients that, on the basis of scientific knowledge, are judged by the FNB to be adequate to meet the known nutrient needs of practically all healthy persons" (NRC, 1989a). We wanted to incorporate new research about the role of diet and specific nutrients in the prevention or reduction of the risk of chronic disease, and we recognized that the RDA alone was not going to be sufficient for all applications.

The DRIs evolved over the next 10 years. Two features in particular set the DRIs apart from the old RDAs. One is that they are based on an explicit functional or physiological criterion or endpoint. The second is that each criterion has a distribution of requirements within the population, assumed to be normal for most nutrients. The changes caused users to begin asking whether different applications required different criteria, and a false sense of confidence about the precision of our understanding of the distribution of intakes and requirements for the population developed. We learned that nutrient requirements are known for only small groups of individuals at one point in time and in one setting, so that standards set for populations are only estimates; and that the goals and process for estimating nutrient requirements differ from those of estimating healthy food patterns to prevent chronic disease and should not be mixed.

DRIs are only as good as the science base on which they are built. Many DRIs stem from metabolic studies, which have limitations as an approach to determining nutrient requirements—healthy people are usually studied, large changes in nutrient intakes are required to overcome homeostatic control of the endpoint (which makes it difficult to get a precise estimate of nutrient requirements), and the studies are expensive. Moreover, available studies often have small sample sizes, which make it difficult to evaluate the true variance in nutrient requirements. Research is needed to give us sensitive, specific measures of nutrient requirements that integrate genetic, environmental, and developmental influences (Figure 2-8).

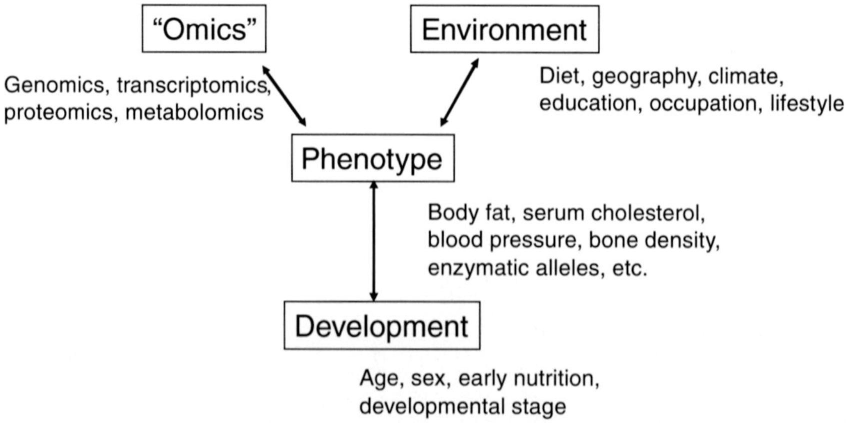

FIGURE 2-8 Research need: Sensitive, specific measures of nutrient requirements that integrate genetic, environmental, and developmental influences.

Several implications of incorporating chronic disease prevention into the DRIs were unanticipated:

- It immediately expanded the science base. We learned that endpoints for reducing the risk of chronic disease have multiple dietary determinants, that the relationship between specific nutrients and disease endpoints varies widely in a population, and that the quantitative information needed to relate nutrients, foods, or food patterns to chronic disease is extremely limited.
- It led to the development of mixed criteria for recommendations and a set of DRIs that gave widely disparate standards among the nutrients. The AIs for calcium and vitamin D are a good example. Both of them were set to reduce the risk of osteoporosis or bone disease, but the endpoints were very different: a desirable calcium retention level for bone for calcium, and the amounts to maintain normal serum 25-hydroxyvitamin D levels for vitamin D.
- The committee expertise needed to incorporate reducing the risk of chronic disease into the dietary guidelines led to the need for a more diverse committee and multiple demands on that committee.
- It led to a new set of recommendations, the AMDRs, defined as the "range of intakes for a particular energy source that is associated with a reduced risk of chronic disease while providing adequate intakes of essential nutrients" (IOM, 2002/2005). However, the AMDRs have flaws that stem from the inadequacy of the research data from which they were derived. Evidence is accumulating that the types of carbohydrates and fat may be more important than the

> total amounts. There was no consideration of the relationship between the intake of sugar and the risk of chronic disease. There were no specific recommendations for dietary fiber or monounsaturated fatty acids. Users were unsure how to apply the qualitative standards for cholesterol, *trans* fatty acids, saturated fats, and sugars. No data were available on upper protein intake levels; the value was actually derived from the AMDRs for carbohydrates and fat, which led to two different sets of protein recommendations.

It is becoming apparent that the DRIs and the *Dietary Guidelines for Americans* have different primary purposes. The DRIs are estimates of nutrient requirements, primarily to prevent nutrient deficiencies and excesses. The dietary guidelines are the basis for the food and nutrition policies in the United States and Canada and for consumer food and nutrition education by the government. In the United States, the DRIs are the science backbone of all nutrition policy. From the DRIs stems the Dietary Guidelines Advisory Committee report that translates the DRIs and additional science linking food, physical activity, and chronic disease into dietary guidelines. From that come two other reports: the *Dietary Guidelines for Americans* policy document and a companion consumer brochure, both prepared by the U.S. Department of Health and Human Services and the USDA. There is also MyPyramid, which develops a food pattern for individuals that stems from both the DRIs and the dietary guidelines.

We need to clearly delineate standards for preventing nutrient deficiency from standards for preventing chronic disease. They have different goals. Different scientific committees with different expertise are needed to address their development, and separate scientific documents are needed to provide the evidence base for the subsequent government and other reviews of these reports. Whether the IOM or a government advisory committee should be responsible for developing these science-based reports is not clear. Also, our research keeps evolving, which leads to questions about which information should be part of the DRIs or the dietary guidelines. For instance, there is science that links the proportion and sources of food and nutrients to individual metabolism. Furthermore, there is emerging research on the link between specific foods in the diet and physiological function.

In conclusion, we, as scientists and users, need to clearly define the DRIs *before* establishing the next process. We should think about establishing EARs and ULs for only one criterion per age/gender group. We need to clearly differentiate the standards for reducing deficiencies of essential nutrients from nutrient and food intakes for reducing chronic disease. We need to keep it simple and, if necessary, have separate reports for different applications to ensure that we are addressing the primary goal in each report. We need to try to find a logical, clear way forward.

EVALUATING EVIDENCE FOR DRI DEVELOPMENT: WHAT ARE THE ISSUES IN APPLYING SYSTEMATIC EVIDENCE-BASED REVIEW APPROACHES TO DRI DEVELOPMENT?

Presenter: Alice H. Lichtenstein

The formal use of SEBRs as part of the DRI development process is intended to supplement, rather than displace the efforts of the IOM study committees and in many cases allow them to focus their limited time on interpreting the available data rather than identifying and collating the information. Therefore, it is helpful to first address general issues about what SEBR can and cannot be expected to do. These are listed in Box 2-4.

Description of SEBR

SEBR comprehensively identifies and tabulates available literature. Appropriately defined questions can supplement traditional approaches to DRI development and increase the consistency of the process. SEBRs are defined by the IOM study committees and other stakeholders. Once the ques-

BOX 2-4
What Systematic Evidence-Based Review (SEBR) Does and Does Not Do

What SEBR Is

- SEBR is one tool for use by the study committee as it deliberates during the Dietary Reference Intake (DRI) process.
- SEBR allows study committee members to focus on the larger picture. Many factors go into making decisions and evaluating the nutritional literature beyond a systematic analysis of the available literature.
- SEBR can offer increased transparency to the DRI process.
- SEBR expands the documentation process in a way that allows for more efficient updating as new data emerge.
- SEBR more precisely identifies research gaps and, therefore, provides a more persuasive argument to target specific funding to close some of those gaps.

What SEBR Is Not

- SEBR does not "automate" the review process and relegate decisions to computer modeling.
- SEBR does not shift the decision-making process from the study committee to the SEBR group.
- SEBR does not diminish the need for expert opinion and scientific judgment.

tions are formulated, a critical step involves refining the questions through discussions with the study committee and other designated individuals to ensure that the intended outcome will be achieved. The various steps in SEBR are described below:

1. *Formulate and refine the question.* The first critical step is to develop the question(s). Individuals requesting the SEBR must clearly and specifically define the question of interest. An example would be: What is the efficacy or association of omega-3 fatty acids in preventing incident CVD outcomes in people without known CVD (primary prevention) and with known CVD (secondary prevention)? During this phase, questions are often refined and clarified by iterative discussions between the group requesting the SEBR and the SEBR panel before the start of the review. Critical components of the questions can be summarized by PICO—population, intervention, comparator, and outcome. In the omega-3 fatty acids example, the population was primary prevention and secondary prevention. The intervention was α-linolenic acid (ALA), eicosapentaenoic acid (EPA), and docosahexaenoic acid (DHA). The comparators were diet and oils containing non-omega-3 fatty acids. The outcome was all-cause mortality, myocardial infarction, stroke, or sudden death.
2. *Develop a search strategy.* The next step is to develop a search strategy. This includes establishing a cutpoint (stop date) for the literature search and identifying relevant databases and other sources of literature (e.g., MEDLINE, PreMEDLINE, Embase, Cochrane Library, Biological Abstracts, Commonwealth Agricultural Bureau of Health, reference lists of reviews/primary articles, suggestions from domain experts).
3. *Identify inclusion/exclusion criteria.* The next step is to develop inclusion/exclusion criteria for the studies. For the example of omega-3 fatty acids and CVD, the criteria were (1) literature published in English; (2) both experimental and observational studies; (3) studies with original outcome data; (4) studies that evaluated all potential sources of omega-3 fatty acids in the diet; (5) studies with at least five subjects; and (6) study duration of at least 1 year.
4. *Retrieve and screen relevant literature.* This is a critical, time-consuming step in the process. For omega-3 fatty acids and primary prevention cohort studies alone, 7,464 studies were identified and their abstracts screened, 768 papers were retrieved (eliminating papers that did not meet the predetermined inclusion criteria), and 118 were identified as being potentially relevant to CVD outcomes. In the end, 39 studies uniquely filled all of the inclusion criteria.
5. *Grade studies.* Studies are graded on methodological quality, appli-

cability, and overall effect. For methodological quality, an A would be given to studies that have the least bias with valid results (e.g., placebo-controlled, blinded, randomized controlled trial); a B study might be a study that is susceptible to some bias, but not sufficient to invalidate the results (e.g., a study where fish oil was given, but not a placebo); and a C study would be one where there was significant bias that could potentially invalidate the results. With respect to applicability, the criteria are determined by the characteristics of the population studied (e.g., [1], sample represents target population, includes both genders, wide age range, and other features of the target population [diet]; [2], sample represents a relevant subgroup of the target population [females], but not the entire population; and [3], sample represents a narrow subgroup of subjects only and is of limited applicability to other subgroups [females between 20 and 30 years of age]). In terms of the overall effect, the categories generally are (1) ++, clinically meaningful, beneficial effect demonstrated; (2) +, clinically meaningful, beneficial trend exists but is not conclusive; (3) 0, clinically meaningful effect not demonstrated or unlikely; and (4) –, harmful effect identified or likely.

6. *Extract/summarize data.* The data are extracted from the literature and summarized in tables. Table 2-2 gives an example of a table generated.
7. *Present report.* The final step is to present the report to the group requesting the SEBR.

Differences Between SEBR for Clinical Medicine and SEBR Needed for DRI Development

The data available to answer clinical medicine questions tend to be more straightforward than issues related to nutrient requirements. For drugs (e.g., statins), in the simplest case, one group is given the active drug and the other group a placebo, and outcomes are assessed. Primary and secondary outcomes are defined ahead of time, and an answer is obtained.

Questions for DRI development tend to be more complex and numerous. The difference from clinical medicine is that an answer must be reached in spite of a high degree of uncertainty. In the medical literature, there would be a cumulative meta-analysis, with one study adding on to the next one until there was enough power to make a determination, then a clinical recommendation would be made. In the case of DRIs, there is a high degree of variability among nutrition studies (e.g., for omega-3 fatty acids, some studies use fish oil supplements, some use fish, etc.; in many cases, the outcome measures or characteristics of the populations were different). The result is that there may be a considerable number of data on a specific nutrient, but cumulatively the data do not lend themselves to merging for increased statistical power.

Special Considerations for Evaluating Evidence for DRI Development

A number of realities about the nature of the currently available evidence in the field of nutrition are important to DRI development overall and specifically to SEBR. Information about the background diet (e.g., fish and omega-3 fatty acids) has already been mentioned. Another is the nutrient status prior to the intervention. If one is supplementing the diet, the effect is going to be different depending on whether the person starts out deficient or nutrient adequate. Examples from the literature include iron and chromium. Adequacy of nutrient stores also needs to be evaluated because someone who is nutrient replete responds differently to supplementation than does someone who lacks adequate stores (e.g., vitamin A). Furthermore, changes in body weight can confound the outcome; these changes commonly occur when the research protocol manipulates fat, carbohydrate, and protein intake and does not adjust the calorie intake.

One needs to consider the bioequivalency of different forms of nutrients. There can also be altered bioavailability due to the co-ingestion of different foods. Other concerns are altered bioavailability from food processing, drug–nutrient interactions, and nutrient–nutrient interactions. Other issues to be considered are altered absorption efficiency due to habitual intake, physiological status, and nonfood contribution of nutrients. Multiple effects of a single nutrient and one nutrient potentially masking the effects of deficiency of a second nutrient are also concerns. Different nutrient bioavailabilities from food and synthetic forms are becoming more important with the high rates of supplementation and nutrients being added to foods or drinks. Finally, we are just learning how to deal with genetic polymorphisms in nutrient metabolisms, as well as how to address nutrient requirements for essential versus nonessential nutrients and energy-containing versus non-energy-containing nutrients. These are likely to require different approaches to the SEBR process.

Implications

SEBRs represent a rigorous process of systematically compiling scientific evidence. They minimize bias through comprehensive and reproducible searches for and selection of articles. They provide rigor by assessing the methodological quality of the included studies and the overall strength of the body of evidence. They enhance transparency through detailed documentation of decisions. They provide useful inputs into program and policy decision-making processes. There is tremendous potential for SEBR to aid

TABLE 2-2 An Example of a Table Summarizing Data Meeting Inclusion Criteria

Table 1. Secondary Prevention Randomized Controlled Trials of Omega-3 Fatty Acid Supplements on Various Cardiovascular Disease Outcomes*

	Omega 3 Fatty Acid		Control			All-Cause Mortality	
Author Year Country (ref)	N	Type Dose Ratio	N	Type Dose	Duration (Year)	Control Group Event Rate (%)	RR 95% CI
ALA vs. EPA + DHA							
Singh 1997 India (16)	120	Mustard Oil ALA 2.9 g/d	118	Non-oil placebo	1	–	nd
	122	EPA + DHA (1:1) 1.8 g/d					
EPA + DHA							
Marchioli 2002 Italy (12)	5665	EPA + DHA (1:2) 0.85 g/d ±Vit E	5658	Control ±Vit E	3.5	9.8	0.79[a] 0.66–0.93
Nilsen 2001 Norway (13)	150	EPA + DHA (1:2) 1.7 g/d	150	Corn oil 1.7 g/d	1.5	7.3	1.0 0.45–202
Leng 1998 Scotland (14)	60	EPA 0.27 g/d	60	Sunflower seed oil 3 g/d	2	5.0	1.0 0.21–4.8
Sacks 1995 U.S. (15)	31	EPA + DHA (3.2) 4.8 g/d	28	Olive oil	2.4	3.6	0.3 0.01–7.1

*Abbreviations: N = number of subjects; RR = relative risk; CI = confidence interval; g/d = grams per day; nd = no data.

[a]Adjusted for main confounders as reported in article.

NOTE: References cited are not included in the reference list at the back of the report. They may be found in the original source.

SOURCE: Wang et al. (2004).

Cardiac Death		Sudden Death		Non-Fatal MI		All Strokes		
Control Group Event Rate (%)	RR 95% CI	Control Group Event Rate (%)	RR 95% CI	Control Group Event Rate (%)	RR 95% CI	Control Group Event Rate (%)	RR 95% CI	Quality
22	0.61 0.34–1.1	6.6	0.25 0.05–1.1	25	0.59[a] 0.35–1.0	–	nd	C
	0.52 0.29–0.95		0.24 0.05–1.1		0.52 0.3–0.9			
5.4	0.65[a] 0.51–0.82	2.7	0.55[a] 0.39–0.77	4.1	0.91 0.70–1.2	1.4	1.2 0.81–1.9	B
5.3	1.0 0.39–2.6	–	nd	10	1.4 0.75–2.6	–	nd	B
–	nd	–	nd	6.7	0.75 0.18–3.2	1.7	3.0 0.32–28 non-fatal	A
3.6	0.3 0.01–7.1	–	nd	7.1	0.45 0.04–4.7	0	2.7 0.12–64	B

future IOM DRI study committees in formulating and revising DRI values. The exact role they will play is difficult to predict. It is likely to vary depending on the specific nutrient of interest. It is hoped that SEBR will be used as a tool to facilitate the process of developing and revising the DRI values.

RISK ASSESSMENT: IS IT A RELEVANT ORGANIZING STRUCTURE?

Presenter: Elizabeth A. Yetley

This presentation considers whether it would be useful to extend the risk assessment framework from its current use as an approach for deriving ULs to future use in deriving the EARs as well as the AIs and AMDRs.

Risk assessment is not a specific methodology, but rather an organizing framework for scientific assessments. It is the scientific arm of a triad of functions that constitute risk analysis (Figure 2-9). Another arm, risk management, reflects those tasks carried out by the users and sponsors of the risk assessment outcomes. There is also a risk communication arm.

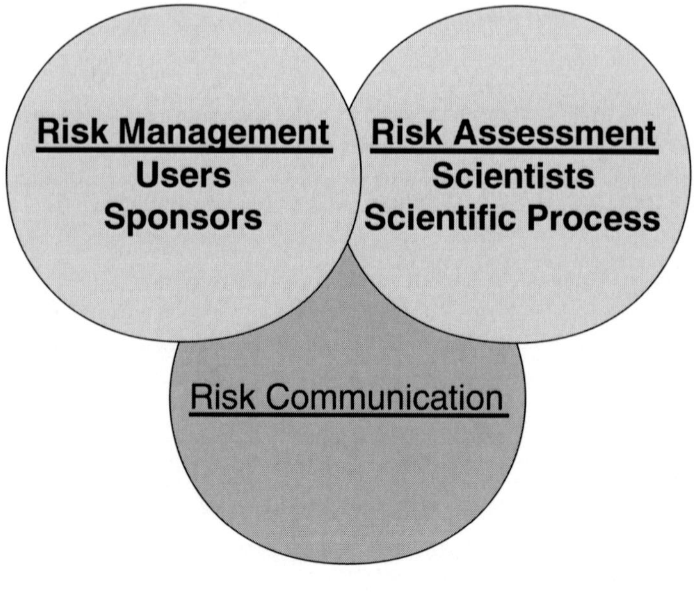

FIGURE 2-9 Risk assessment as part of the risk analysis triad.

Risk Assessment: A Systematic Process

The conceptual underpinnings of risk assessment stem from a 1983 National Research Council (NRC) publication on how scientific deliberations should be organized to assess risk in a manner that meets user/sponsor needs while maintaining the scientific integrity of the assessment (NRC, 1983). The risk assessment framework includes guidance on managing interactions between the sponsors or users of the risk assessment and the scientists conducting the risk assessment.

The current ULs issued by the IOM used a nutrient risk assessment framework. The United Kingdom and European Commission (EC) used a similar approach to derive their upper level reference values for nutrients. The approach evolved from applications originally developed for chemical contaminants and subsequently modified for application to microbial pathogens.

Several characteristics associated with risk assessment are described in the 1983 and subsequent NRC documents. First, no decision is often not a viable option from the perspective of protecting public health. It was deemed better to have an informed decision based on the best scientific expertise, even if not perfect, than no decision that by default provided no guidance for evaluating the status quo. This is often also true for essential nutrients.

Second, as it developed the risk assessment framework, the NRC recognized that it would usually have incomplete data, and that uncertainties would need to be dealt with through documentation and use of expert scientific judgment. This need for dealing with evidentiary uncertainties is also true in deriving nutrient reference values.

Third, the NRC focused on the needs of users/sponsors in developing the risk assessment framework. Users need science that addresses their information needs and is also presented in a manner that allows them to readily integrate results into program or policy initiatives. This requires emphasis on a mutual understanding between sponsors and risk assessors of user information needs and on transparency and documentation of the series of decisions made in a risk assessment. At the same time, the NRC wanted to protect the scientific reviewers from undue stakeholder pressure by ensuring independence of the scientific evaluations.

The NRC developed a systematic process (Figure 2-10) that goes through a series of evaluations and decision steps. Within each of these steps is an articulation of the basis and rationale for each type of decision. This helps the user understand the rationale for decisions and therefore enhances usability for a broad range of applications.

One component of the risk assessment process defines the rules of engagement between the sponsors/users of the risk assessment and

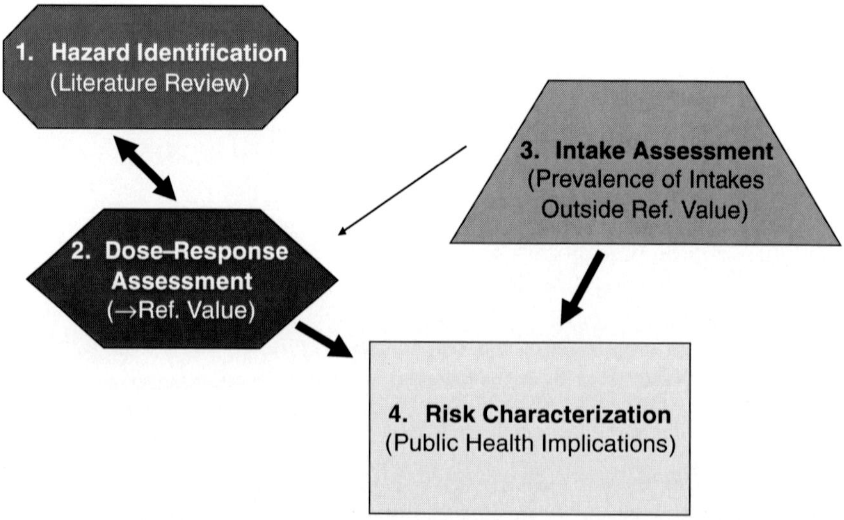

FIGURE 2-10 Steps in risk assessment.
NOTE: Dark arrows represent major pathways; light arrow represents a pathway used less often, but feasible.

the scientists conducting the actual assessment (Figure 2-11). The risk assessment itself is pictured in the central area of this figure. The scientific assessment would be equivalent to a DRI study committee. The risk assessment process differentiates between the roles and responsibilities of the risk assessment study committee and the sponsors who have requested the risk assessment. There is also emphasis on ensuring that the results of the scientific assessment are presented in a manner that enhances their usefulness to sponsors and other users.

The sponsors are responsible for defining questions that need to be addressed before the risk assessment process starts. This is called *problem formulation*. For example, in the case of the DRIs, the problem formulation statement could specify the populations to be covered (e.g., healthy versus general populations, individuals versus groups). Public input may be solicited during this process. The sponsors, with or without public input, would identify for the risk assessors the questions and issues that the risk assessment should address. At this point, there often needs to be dialogue between the sponsors and risk assessors to ensure that the scientists understand the information needs of the users or sponsors and the risk assessors have the opportunity to suggest revision and clarification of the problem formulation questions, if needed.

As the scientific assessment is completed, its results need to be expressed

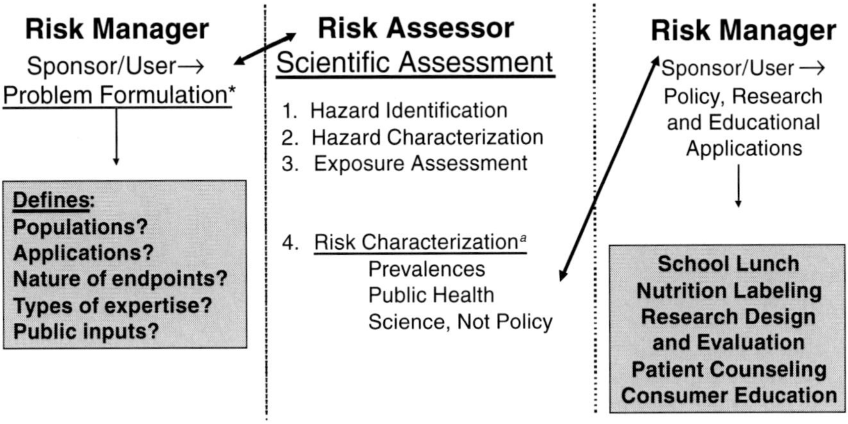

FIGURE 2-11 Risk analysis: How to increase usability while maintaining scientific integrity?
[a]Critical interaction components.

in a manner that will enhance its usefulness to sponsors and users. This includes full development of the last step of the risk assessment framework, the so-called *risk characterization* step. For example, the committee would describe the nature of the risks associated with inadequate or excessive intakes and the percentage of the population exceeding the UL or failing to meet the reference value of adequacy. The risk characterization step also describes the public health implications of these deviations. For example, if 20 percent of children are consuming some nutrient above the UL, is that likely a serious public health problem? However, the risk assessment should not stray into policy recommendations in the risk characterization step because this could cast doubt on the scientific independence and integrity of the risk assessment. The risk assessors would indicate that a certain segment of the population appears to have a public health risk based on the intake and the reference value and would describe the public health implications of this deviation. However, the risk assessment study committee would not recommend public health actions. For example, it would be inappropriate for the risk assessment study committee to recommend that its evaluation suggests the need to fortify the national food supply or change school lunch standards.

These results of the risk assessment need to be described and documented in a form that enhances usability to the risk manager, the sponsor, and the users. Interestingly, the two steps designed to maximize the usefulness of a risk assessment review—the problem formulation and the risk characterization steps—are probably the two steps in the process that have been least developed, although they are critical for bridging the gap between sponsors and scientists.

In brief, risk assessment is used to derive science-based evaluations. It is recognized that the evaluations may have public health implications, although they often need to be made with evidentiary uncertainties. Risk assessment focuses on user needs by clarifying the information needs, documenting decisions, and describing the public health implications of population deviations from reference values in the risk characterization step. At the same time, the vertical lines between the groups in Figure 2-11 are designed to protect the scientific integrity of the scientific assessment once user needs are understood by the scientific assessors. Nevertheless, the risk assessors may need to reinitiate dialogue with the sponsors/users occasionally during the process because unanticipated challenges may require further clarification.

Applying the Risk Assessment Framework to Indicators of Adequacy

When a risk assessment framework is applied to a new discipline, the basic conceptual framework generally stays the same, but some adaptations and redefinitions of terms are often needed to ensure relevance to the new disciplinary application. This applies to the use of risk assessment frameworks for indicators of nutrient adequacy.

One benefit of a risk assessment framework is that the usability of the reports is enhanced because of the focus on meeting user needs. Another benefit is the enhancement of the quality of the scientific assessments. If, for example, study committees used the same organizing framework to derive both adequate and excessive intakes, it would allow side-by-side comparisons of the evaluations and decisions for both as the scientists go through the decision-making processes. This could help in identifying unintended inconsistencies between decisions resulting from evaluations of adequate and excessive intakes. It could also allow concurrent examination of prevalences above the UL and below the reference value for adequacy within and across life stage groups for a given nutrient. This is potentially important to users who are frequently faced with balancing the conflicting needs of low-intake consumers with the potential for excessive intakes among other consumers.

In the classic risk curve for the DRIs (Figure 2-12), a two-tailed risk curve illustrates the increased risks of adverse effects associated with both excessive intakes and inadequate intakes. The general risk assessment community is already moving from a one-tailed evaluation of adverse effects associated with toxic levels of intake to concurrently looking at a two-tailed risk curve that examines the potential for unintended risks associated with actions that would reduce access to foods containing toxic contaminants.

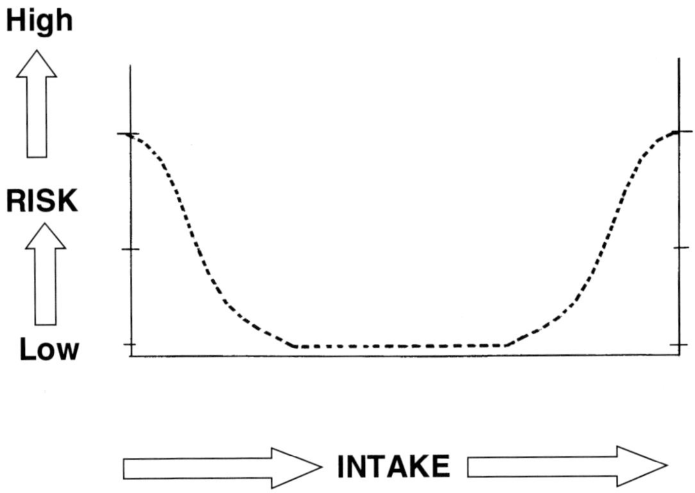

FIGURE 2-12 Levels of risk associated with change in intake from inadequate to excessive intake.

For example, when considering the risks associated with contaminants in fish or fruits and vegetables, public health considerations may also benefit from a concurrent evaluation of risks associated with decreased consumption of these foods. These two-tailed evaluations of risk are called *risk/risk assessments* or *risk/benefit assessments*. Either one of these terminologies and concepts could apply to nutrients because risks are associated with both inadequate and excessive intakes.

The first step in a risk assessment model is *hazard identification* (see Figure 2-10). "Hazard" is defined by the World Health Organization as "an inherent property of the substance that causes harm" (IPCS, 2004). This clearly relates to harm associated with excessive nutrient intakes. However, it does not apply to risks associated with nutrient inadequacies because an adverse effect associated with an inadequate nutrient intake is not due to an inherent property of that nutrient. It is simply because the nutrient is lacking. Thus, the terminology would need to be revised to apply risk assessment approaches to derivation of reference values for nutrient adequacy. For example, for nutrient risk assessment purposes, the terminology might be changed to "identification of indicators of adequacy (or inadequacy) and hazard."

As indicated, the systematic risk assessment approach goes through a series of four steps, each with a series of mini-steps and decisions. For each step, the decisions made for ULs are generally similar to the decisions made for adequacy. Thus, the risk assessment framework could easily be adapted to both sides of the equation.

The first step, hazard identification, is basically the literature review. In general, the nature of the questions for the indicators of hazard for the ULs is the same as that for the indicators of adequacy (e.g., intake/biomarker, biomarker/effect, intake/effect relationships). Both evaluations are focused on identifying dose–response effects and factors that affect dose–response, and both evaluations need this information across a range of life stage groups.

The second step, the *dose–response assessment*, is the step where the reference values (e.g., EARs, ULs) are derived. In deriving the DRIs, a threshold model of dose–response was assumed for both adequate and excessive intakes. In both cases, the study committees frequently lacked good dose–response data. In the case of the UL, if they lacked dose–response data, they used a NOAEL or a LOAEL as the basis for deriving the UL. In the case of the adequacy evaluations, if the study committees lacked dose–response data, they derived an AI. In both cases, study committees preferred full distributions of dose–response data: On the UL side, it is called the benchmark dose; on the adequacy side, the EAR/RDA distribution curves. For both, there have been questions about whether a threshold model always works.

In terms of adjustments to the dose–response relationship, bioavailability and bioequivalency issues relate to risks associated with both inadequate and excessive intakes. For both EAR or AI and UL, the traditional adjustments for bioavailability or bioequivalency for adequate intakes may lack relevance to the UL. For example, the EAR/RDA for iron adjusts for differences in bioavailability from food sources based on dietary intakes of heme/nonheme iron sources. However, with the increasing use of fortified foods and dietary supplements, a more appropriate bioavailability adjustment might be a bioequivalency type of adjustment similar to that used for retinol equivalents. Study committees would likely notice these potential incompatibilities if the evaluations for both adequate and toxic intakes were compared in a side-by-side risk assessment framework. Additionally, the same methodological biases in the studies used to evaluate risks associated with both inadequate and excessive intakes likely occur, so a consistent framework for analyzing both makes sense.

Uncertainty assessments are a critical component in the dose–response assessment step of a risk assessment framework. Derivations of reference values for both inadequate and excessive intakes must deal with uncertainties in the available evidence and describe the nature and seriousness of those uncertainties in their texts. In some cases, an uncertainty factor is used to lower the observed effect level to give a UL. The use of uncertainty factors was relatively rare in deriving reference values for adequacy.

However, at least in the case of vitamin D, the study committee multiplied the observed intake of vitamin D by 2 to raise the AI above the observed dose–response relationship to account for uncertainties in background exposure to sunlight and study design inadequacies. Whether dealing with risk associated with inadequate intakes or excessive intakes, uncertainties need to be documented and are generally dealt with in a manner that errs on the side of public health protection. A risk assessment framework identifies the need to deal with uncertainties in the available evidence, but does not specify a methodology to deal with the identified uncertainties. This allows maximum flexibility in applying this organizing framework to different situations.

Establishing reference values for both inadequate and excessive intakes also often involves extrapolations from a studied group (e.g., adults) to an unstudied group (e.g., children) because data may be available for some, but not all, of the life stage groups for which DRIs are established. The default for extrapolations for ULs was reference body weight. The default for EARs/AIs was metabolic body weight. There is no acknowledgment of, nor justification for, the use of different defaults for these two types of reference values. If there was a side-by-side common risk assessment framework, these types of differences would likely be noted and either justified or changed.

The third step, *intake (or exposure) assessment*, uses population-based intake data to estimate the prevalence of intakes above or below the reference values. Biomarkers of nutrient status, when available, can also be used to estimate prevalence of inadequate or excessive exposures. The same analysis is often used for both types of reference values. The fourth step is *risk characterization*, which is the most important step from a user perspective. This is where the public health consequences of not meeting an EAR/RDA or AI and exceeding a UL are discussed. Deviations from reference values for special groups are also described in this section.

Implications

An advantage of using a risk assessment framework is that the science of risk assessment has been moving forward. The DRI development process can benefit from these efforts. For example, risk assessors increasingly have been using probabilistic models to move from qualitative to quantitative risk assessments. They have been working to establish better defined criteria for dealing with different types and sources of uncertainty. They are starting to use statistical models to simulate dose–response curves from multiple studies that individually lack sufficient data to produce a dose–response curve. They are also learning how to adjust coefficients of variability to account for altered dose–response curves associated with polymorphisms that alter nutrient requirements or toxicity among population groups.

In summary, the risk assessment organizing framework is probably relevant to the development of reference values related to nutrient adequacy. It provides a systematic delineation of decision steps that enhances transparency and therefore increases usability. A risk assessment organizing framework could help coordinate the decisions related to adequacy with those related to excessive intake, thus reducing the likelihood of unintended inconsistencies or consequences that create challenges for users. The risk assessment framework offers the flexibility to tailor the approach to different types of applications without losing the benefits of the organizing framework. It emphasizes enhanced documentation and transparency and takes advantage of evolving scientific tools.

DISCUSSION: SYSTEMATIC EVIDENCE-BASED REVIEW; RISK ASSESSMENT

Discussant: Sanford Miller

The session moderator, Dr. Stephanie Atkinson, introduced the discussant and invited him to offer an opening remark.

Discussant Opening Remarks

Dr. Miller opened with the general observation that although it seems we are asking the same questions from years ago, we are learning to ask better questions. He noted that it is not surprising that study committees appeared to derive their own approach to the problems they faced given the lack of experience, structure, or formal guidance when the DRI process began. He suggested that risk assessment and SEBR together provide an excellent framework to organize the process and the questions to be addressed as well as structure to allow transparency on how conclusions were reached or the rationale for why a modified approach was used by a particular committee. Dr. Miller then focused on the nature of the relationship between risk and dose–response. The two risk curves associated with nutrients are composed of families of curves, and in turn represent the components of the metabolic regulatory process for absorption and excretion. If there is uncertainty around the curves, they will overlap, suggesting the nutrient is unsafe at the same time that it is required. For this reason, it is critical to carry out basic research focused on the process by which a nutrient is used and regulated in order to reduce the level of uncertainty.

General Discussion

Drs. Lichtenstein and Yetley joined the discussant on the dais, and a brief group discussion took place. They agreed that the DRI approach was a vast improvement over the previous RDA approach, and that these reference values should improve as science evolves and experience is gained. It was noted that the past 10 years of experience have led to a greater understanding of the breadth of potential uses for the DRIs, which is important when making future revisions.

When the discussion was opened to the workshop audience, comments were offered on several topics, including SEBRs and risk assessment.

SEBRs

One audience member suggested that although SEBRs are important, they may have limitations. They add time as well as cost because the individuals performing the SEBRs are not likely to be volunteers. Furthermore, the development of DRIs inherently requires scientific judgment, which requires a wide range of information about the nutrient. The SEBR focuses on a limited group of questions. A participant responded that individuals commissioned to generate the SEBRs are not charged with offering the scientific judgment necessary for deriving the DRIs. The advantage of study committees working with an evidence-based practice group is that once the relevant questions for the targeted review are defined by the study committee, the practice group can examine the evidence in an objective manner. Database limitations for most questions and the number of questions that can be addressed for each nutrient mean that, ultimately, the judgment of the study committee is required. Thus, SEBRs would not be used to derive DRI reference values; rather, they would be used as one source of data for deriving the reference values.

A commenter remarked that SEBRs can be carried out by either paid panel members or unpaid volunteers who "work outside of their day jobs." She then inquired about the professional expertise needed for these SEBR panels as opposed to the DRI study committees, and about the rewards for unpaid volunteers. A participant responded that SEBRs are not work carried out in spare time. They must be done in a consistent manner and require considerable amounts of time, focus, and resources. Regarding why people are willing to take part in these activities, the discussant suggested that those who believe nutrition is fundamental to reducing the risk of disease will feel a responsibility to participate.

Another participant noted that SEBRs do not replace the need for a DRI study committee, but instead serve as a tool to help document, collate, and synthesize the scientific evidence. This tool could lessen the burden of the study committees and allow them to focus on the challenges of defining DRI values. Nor is the SEBR competing with the risk assessment framework. The first step in the risk assessment approach is a literature review; if SEBR were used, it would feed into the larger risk assessment activities. For this type of review, the study committee would help to define the inclusion/exclusion criteria for the literature, the endpoints to be reviewed, and other considerations. In fact, the use of a risk assessment framework to help organize activities could make more efficient use of staff time and volunteer time. Outside help with the literature review could relieve the study committee members—who are volunteers—of the burden and allow them more time for other needed deliberations.

Additional Comments

Other comments on risk assessment as well as SEBRs included the following: the risk assessment framework requires that uncertainties be dealt with, but the methods to be used are unspecified and can be determined case by case; the process of SEBR is robust and not limited to a particular kind of study design; SEBRs can expose gaps in knowledge; and not every question addressed by a study committee would require SEBR.

With respect to the DRI framework, an audience member suggested that the EAR/RDA is related to measures of central tendency whereas the UL is not. He postulated that the UL is more analogous to the AI in that the AI is above the amount needed while the UL is below the amount to be avoided. Furthermore, it would be possible to define a level for adequacy in a manner similar to that for developing ULs. He said the value of doing so and whether the data would support it are important discussion points.

One person noted that death from disease had not been mentioned as a marker for chronic disease risk in the DRI process, even though there may be a reduction in death from disease associated with some nutrients. A participant responded that in the case of DRIs, death is probably not a preferable measure as compared with appropriately validated biomarkers for the advent of the disease state. The final comment of the discussion related to the value of testing intake recommendations as they are being developed. An audience member used the example of a reasonableness check for iron recommendations and its ability to better inform the process and thus lead to better outcomes.

PANEL DISCUSSION: IN WHAT WAYS COULD THE CONCEPTUAL FRAMEWORK BE ENHANCED?

Panel Members: Cutberto Garza, Mary L'Abbé, Irwin Rosenberg, Barbara Stoecker
(later joined by Janet King and George Beaton)

The session moderator, Dr. Stephanie Atkinson, introduced the panel members and began the discussion by asking each panelist to offer an opening remark.

Panelist Opening Remarks

Dr. Garza highlighted three conclusions that he drew from the day's discussions. One was the need to "keep it simple." At the same time, he emphasized that we need to be more sophisticated so that finding the simple solution does not result in the wrong answer. He cautioned that this sophistication would be needed especially if we move from focusing on preventing deficiency and diet-related diseases to focusing on enhanced performance. In this case, it may not be reasonable to expect (simplistically) that the frameworks best equipped to deal with deficiency, chronic disease, and enhanced performance will be the same. Second, he suggested harmonizing approaches for deriving the EAR and the UL to allow greater transparency globally and to enhance the rigor of the process, regardless of the degree of precision needed. He made the analogy to the hazard analysis and critical control points approach used to ensure food safety, where control points are identified in the process. Third, the dynamic nature of the field needs to be recognized, and the DRI framework should reflect that dynamism. He emphasized that the dynamism will dictate the type of evidence collected, the criteria for deciding when the numbers need to be revised, and even the format in which the DRIs are published.

Dr. L'Abbé touched on the importance of the underlying theme of transparency, specifically from the perspective of a government agency that uses the DRIs in a number of applications (e.g., food fortification, product evaluations, standards setting). She then underscored the need for DRIs to be relevant to public health risk. Finally, she pointed out that to apply the values effectively, regulators and government agencies need to understand the process and the approach to decision making used by the study committees. Conversely, sponsoring government organizations bear the responsibility of defining the general questions to be answered through the process of DRI development if the end result is to be useful.

Dr. Rosenberg remarked that realizing the conceptual framework that we seek holds considerable challenges. Essential to our success will be a consensus on the overall goal of the DRIs. If the goal is to sustain the health of the North American population, it must be recognized that this is not the only sustaining pillar of public health. Others include the dietary guidelines and relevant reports from the Office of the Surgeon General. He cautioned that there is risk in using DRI values to cross into dietary guidelines; in turn, this can spawn some confusing concepts, such as semi-quantitative AMDRs for nonessential nutrients. Moreover, despite recent assertions to "change" our paradigm to include chronic disease prevention, the goals for the reference values issued by the NRC and then the IOM have remained remarkably stable since 1941. These dietary recommendations have always been more than minimal allowances and have by implication included prevention of chronic disease as part of the definition of good health maintenance. As a last point, Dr. Rosenberg addressed the issue of multiple endpoints. The current approach for DRIs uses different endpoints for different population groups (children, pregnant women, sometimes the elderly). However, the argument that study committees should issue, and users choose, different endpoints for the *same* group would lead to misunderstandings and undermine the integrity of the process.

Dr. Stoecker noted that the public has a false sense of confidence about the knowledge available for setting the DRIs. She commented that, of course, the data on many nutrients are scarce. She noted particularly that dose–response data at intakes near the probable EAR are needed, but may be difficult to obtain. Dr. Stoecker supported the use of the risk assessment approach as an organizing structure and agreed that SEBRs organize, document, and encourage transparency of the process. Furthermore, she suggested that nutrient requirements and chronic disease prevention could be dealt with in separate reports because of the multiple factors that affect the chronic disease endpoints compared with the nutrient requirement endpoints. She agreed with the conclusion that a single endpoint should be used for age/gender groups.

General Discussion

Cross-Panel Discussion

The cross-panel discussion covered several topics. It began with several comments on endpoints.

Endpoints One participant suggested that a variety of endpoints are typically expected to be considered using all of the emerging science. Then, based on clear criteria, the endpoint to serve as the basis for the reference value would be selected. However, the criteria for selecting an endpoint have not been clear. Rather, endpoints seem to have been determined primarily on the basis of data availability, which does not necessarily reflect health significance. Another participant noted that she found comfort in discovering that often the same "ballpark" value could be derived using any one of a number of endpoints. She also suggested that chronic disease protection might result in a higher reference value. Another panel member responded that although there is likely to be substantial variation among nutrients, it is not clear that the amount of a nutrient required to reduce the risk of chronic disease is necessarily going to be higher than the amount to achieve some other endpoint.

In response to a comment on the apparent inconsistency of the severity or the public health significance of the endpoints used for ULs, a participant noted that in setting food fortification limits in Canada, information in the text of the DRI reports was used to elucidate the relative severity of the adverse effect and the margin of safety between the RDA/AI and UL. She further noted that challenges were presented by nutrient substances such as saturated fats or *trans* fats, which do not fit the classic threshold models for a UL. Another participant said we often fail to look at "population-attributable benefits" and asked: Is there a situation in which a UL could be set too low because a population-attributable benefit of greater public health significance than the mild physiological discomfort used to establish the UL was not taken into account? One participant suggested that from his experience with ULs, the problem was the need to adapt a toxicological model appropriately for nutrient considerations, but there is nonetheless considerable opportunity for parallelism. He asserted that consistency is important, but should not always be expected because key considerations may vary by nutrient and need to be addressed in different ways. Another participant commented that we should not be hobbled by consistency, and it should never preempt scientific rigor.

Precision A panel member pointed out that if we are clear about the various uses of the reference values, we can better assess the degree of precision needed. That is, we are too often driven by an obsession for the precision that our training requires, but that the use does not demand. Assuming this hurdle is passed, the "biology of the nutrient" is the next component to consider, because the inability to specify the biological workings of the nutrient would be limiting in establishing meaningful reference values. From this point, the instruments and organizing approaches we have at our disposal to address the tasks become the focus.

Ranking evidence In response to the suggestion that ranking evidence was a considerable leap for study committees, a panel member said study committee members did discuss the criteria for judging the evidence and did reject studies, so ranking evidence was not necessarily a challenge. Rather, a major shortcoming in the past was the failure to document discussions and the decision-making process.

Open Discussion

The cross-panel discussion was followed by a wide-ranging discussion between audience members and the panel on topics such as the appropriateness of AIs, limited data, updating DRIs, and the interest in harmonization. An audience member suggested a focus on terminology, asking participants to consider terms such as "critical effect," as used in toxicology.

Appropriateness of AIs A participant asked if the AI system should be retained and, if not, whether an EAR should be approximated in some fashion. Another said it should be eliminated. An audience member asked whether any advances were experienced in either the application or communication of the DRIs by incorporating the AI, and whether the AI belongs within the DRI framework. For some nutrients, an EAR could have been derived if a physiological function of the nutrient had been used as the criterion rather than a chronic disease endpoint. A participant reported that study committees were dissatisfied with the advent of the AI because they had been given the task to derive a value based on an endpoint, and they did not feel confident that an AI was appropriate given this charge. In terms of the evolution of the AI, another participant noted that it was used initially to describe recommended intakes for infants, specifically infants that are exclusively breastfed and are thus the easiest population for which to measure dietary intake. In the case of the breastfed baby, the AI values are probably more solid than for most other groups.

Limited data The discussion turned to considering the "no decision is not an option" component of DRI development. One participant expressed concern that numbers developed in the face of limited data appear to take on the same level of significance and credibility as other more well-founded reference values. He recalled that when AIs were first discussed, there was some mention of adding table footnotes or color codes or using faint print as a way of communicating the level of confidence associated with the numbers. He expressed his opinion that, in the end, it seemed that once a value was listed in a table, no one read the footnotes or went back to read the reports. Another participant commented that we should be clear on how the values will be used, then make decisions with respect to how the data are presented in the table.

In the case of ULs, there was considerable agreement on the need to create a reference value if the data supported doing so because in the absence of such a value, various misinterpretations have occurred, including the conclusion that there is no risk. One participant suggested that it would be helpful to set out explicitly the disagreements that occur, indicating the level of confidence in the values offered. Others agreed that the approach used to arrive at the ULs should be described more explicitly. There was concern that reaching far beyond the available data to establish a UL is undesirable; there is a distinction between inadequate data and limited data.

Other discussion focused on obtaining data on the distribution of requirements in order to enhance the DRI process. One participant suggested that although it would be prohibitively expensive to explore the nature of the distributions in detail, we need at least general information such as breadth and skew. In turn, the study committees should be charged with providing such information. Users of the DRIs may prefer to use a value along the distribution curve rather than the RDA for a certain application. Furthermore, it was suggested that the variance of the requirement distribution is not critical to the application except when individuals are concerned. The default CV of 10 or perhaps 20 percent may be adequate in terms of needed precision for practitioners.

Updating DRIs With respect to the future DRI process, one participant asked how we can ensure corrections are made to "mistakes" recognized after the study committee disbands. Another participant responded that the framework should address this, not only to correct mistakes, but also because new science will inevitably dictate changes in reference values. A panel member further suggested that the ability to issue DRIs in the format of a downloadable loose-leaf-type notebook is important so that any needed changes can be addressed and made accessible without having to engage in an entire review. It was noted that using SEBR as part of the process would facilitate any updating.

One participant remarked that when some of the research questions listed in the report were addressed, this information could serve as a trigger for review. Another participant countered that a nutrient should be reviewed when the nature of the outcome will be important to public health. An audience member asked about the nature of guidelines for prioritizing the nutrients to be updated. A panel member suggested that setting criteria for revision and setting criteria for prioritization were different issues and that criteria for revision would be addressed later in the workshop.

Harmonization A discussion clarified that the harmonization referred to by Dr. Garza was a harmonization of *approach* for deriving the numbers rather than of the specific reference values. Dr. Garza suggested that the only values needed globally are the equivalent of the EAR and the UL, and presumably good science could be brought to bear on deriving these. If the approach for deriving these could be harmonized, different countries could, within the context of their own public health protection considerations, derive their own relevant reference values.

Also with respect to harmonization, an audience member suggested that more international expertise should be included in the DRI process so countries could learn from each other, share information, and reduce costs. The benefit to countries not able to mount such a process was also highlighted. Another participant remarked that the EC was working to create a framework for nutrient reference values and harmonize nutrient recommendations across Europe. He suggested benefits in collaboration given the apparent lack of an overall framework for the DRIs. While recognizing the value of collaboration, participants disagreed with the intimation that there was no overall framework for the North American DRI process. Rather, the day's discussions demonstrated that there was a framework, but it may not have been structured or communicated as well as possible.

3

Criteria for Scientific Decision Making: Session 2[1]

During the planning phase of the workshop, Session 2 participants were requested to take into account the same general questions asked of Session 1 participants (see Box 2-1). However, they were specifically asked to each address different decision-making criteria important to the development of Dietary Reference Intakes (DRIs). These criteria are components of the "road map" for DRI development as described earlier (see Chapter 1). General questions were asked of each participant: Can we provide more specific guidance to study committees on scientific decision making to help clarify the concepts and tasks and to promote consistency across study committees? Can we provide guidance to study committees on the use of scientific judgment in the face of limited data that would allow such judgment to be more transparent and better documented?

The second session was moderated by Dr. Robert Russell of Tufts University. Dr. Irwin Rosenberg, also of Tufts University and former chair of the Food and Nutrition Board (FNB), opened the session with a talk on the selection of endpoints. Dr. Susan Taylor Mayne, a professor in the Division of Chronic Disease Epidemiology at the Yale School of Public Health, then spoke on the options available in the face of limited dose–response data.

Dr. Stephanie Atkinson, a professor in the Department of Pediatrics at McMaster University, discussed the challenges in addressing extrapolations and interpolations for unstudied groups. Dr. Hildegard Przyrembel, from

[1]This chapter is an edited version of remarks presented by Drs. Rosenberg, Mayne, Atkinson, Przyrembel, Subar, and Garza at the workshop. Discussions are composites of input from various discussants, presenters, moderators, panelists, and audience members.

the Federal Institute for Risk Assessment in Berlin, spoke on the challenges in addressing adjustment for data uncertainty.

Dr. Amy Subar, a research nutritionist at the National Cancer Institute (NCI), gave a presentation on the implications of estimating dietary intake for DRI development. Finally, Dr. Cutberto Garza, provost and dean of faculties at Boston College and former chair of the FNB, closed the session with some highlights of physiological, genomic, and environmental factors that are important to the DRI process. Discussions and comment periods were held throughout the session.

SELECTING ENDPOINTS: WHAT ARE THE ISSUES AND WHAT ARE THE OPTIONS FOR CRITERIA?

Presenter: Irwin Rosenberg

Endpoints play a pivotal role in the DRI process. They are the skeletal structure on which the Estimated Average Requirements (EARs) and tolerable upper intake levels (ULs) are draped. In essence, they are an expression of the targets or goals of the DRI development process. They should be related to quantifiable or measurable attributes that relate to the overall public health goal of the project. The key concerns from the perspective of selecting endpoints are "adequacy for what ends?" with respect to the EARs and "adverse effects as reflected by what?" for the ULs.

Experience in Selecting Endpoints

Since the 1941 National Research Council (NRC) report (1941), the selection of endpoints for nutrient reference values has evolved in response to changes in nutrition science. These advances sometimes revealed associations between an endpoint and diet and at other times identified possible endpoints through better understanding of metabolic and physiological states. Moreover, approaches for endpoint selection have been variable across the study committees responsible for the reference values. This is to be expected, given the differences in the biology and functions of essential nutrients.

Throughout the experience of developing reference values, limited data have often precluded the identification of the most appropriate endpoint for any given age/gender category. This situation in many cases results in the need to extrapolate knowledge about the endpoint used for one group that is better studied (e.g., adults) to a less well-studied group (e.g., children). This is one area of work that needs further exploration (see presentation by Dr. Atkinson in this chapter).

Importantly, limited data on dose–response relationships have always made it difficult to compare, consider, and prioritize endpoints for the purposes of establishing a reference value. The reliance on studies that examined dose ranges not relevant to adequacy considerations is not a desirable solution. Although meta-analysis studies offer some promises and newer strategies are being developed to deal with limited data (see next presentation by Dr. Mayne), the ideal situation is to have better data.

As we have experienced during the past 10 years, endpoints for specific chronic diseases are especially challenging. Although it is desirable to have chronic diseases as the targets for our requirements and thereby reference values, that was possible in only a few instances. However, we need to recognize that the use of chronic disease as a basis for reference values is not a new paradigm.

Throughout the history of the Recommended Dietary Allowances (RDAs), chronic disease has been an implicit part of trying to set reference values that were above those necessary to prevent deficiency. The idea of achieving the health of the population—and thereby including the risk of chronic disease as an endpoint—has always been present at some level within the process. Whether this can be done explicitly, as was done for some macronutrients in the last series, will require further discussion.

Finally, I would like to make a few quick points on the lessons we are now considering. First, one question raised to workshop participants is the issuing of multiple reference values based on multiple endpoints for a single age/gender group. This is not the issue of study committees considering multiple endpoints before they select one to serve as the basis for reference values, but assigning them the task of issuing values for the various endpoints. Specifying multiple endpoints for a nutrient within a given age/gender group is not useful or appropriate; in fact, it could be very confusing. Rather, a single endpoint for the age/gender group should be selected. Second, the question of whether reference values—EARs, RDA, ULs—are to address essential nutrients only or be expanded to nonessential nutrients, such as fiber and carbohydrate, needs to be considered, particularly in light of our understanding about the interface between the DRI process and food-based dietary guidance.

Selecting Endpoints

In the past, a number of endpoint types have served as the basis for reference values. These have included clinical signs, measures of developmental abnormalities in children, biochemical measures, balance study outcomes, body pool measures, functional measures, and measures of chronic disease risk.

A 1994 Institute of Medicine (IOM) document (1994) lists the types of evidence that have been used in establishing RDAs. These include

- biochemical measurements that assess the degree of tissue saturation or adequacy of molecular function in relation to nutrient intake;
- nutrient depletion and repletion studies in which subjects are maintained on diets containing marginally low or deficient levels of a nutrient, and then the deficit is corrected with measured amounts of that nutrient;
- balance studies that measure nutrient status in relation to intake;
- epidemiological observations of populations in which the clinical consequences of nutrient deficiencies are corrected by dietary improvement;
- extrapolation from animal experiments (although applying animal data to human studies is difficult); and
- nutrient intakes observed in apparently normal, healthy people, which was one way of arriving at an Adequate Intake (AI).

If one views the stages of nutrient insufficiency as a series or cascade of events that describe the temporal sequence of deficiency of, for example, a given vitamin, the initial stages could be called "subclinical deficiency," or findings that would occur before symptoms or signs of disease (e.g., low circulating levels of nutrient, decreased tissue levels or desaturation of body pools, and metabolic disruption). The more advanced stages of deficiency, which could be called "clinical deficiency," encompass the symptoms and/or signs of disease (e.g., reversible changes in the skin and irreversible changes or cell death).

An emerging area important to the criteria for selecting an endpoint is the ability to use a biomarker or surrogate as an endpoint reflective of the functional or clinical response of interest. I will conclude my remarks by reviewing the case of vitamin D.

The vitamin D case is an interesting example because circulating levels of 25-hydroxyvitamin D have been shown to be related to intake of vitamin D. Although this is complicated by synthesis in the skin as a result of sun exposure, it is generally a good measure of absorption of vitamin D. However, data may be emerging that relate levels of 25-hydroxyvitamin D to measures of bone density, skeletal disease risk (as in the case of osteoporotic fracture), and other disease risk (as in the case of extraskeletal cancer and even some immune dysfunctions). Moreover, evidence that 25-hydroxyvitamin D is related to the absorption fraction for calcium suggests that 25-hydroxyvitamin D values have the potential to serve as a target endpoint for an important function and may demonstrate certain convergence with other observations—for example, a lower risk of several kinds of cancer, at least in some intervention studies.

A regression meta-analysis reported by Bischoff-Ferrari et al. (2005) shows that in a number of studies, a significant decrease in relative risk

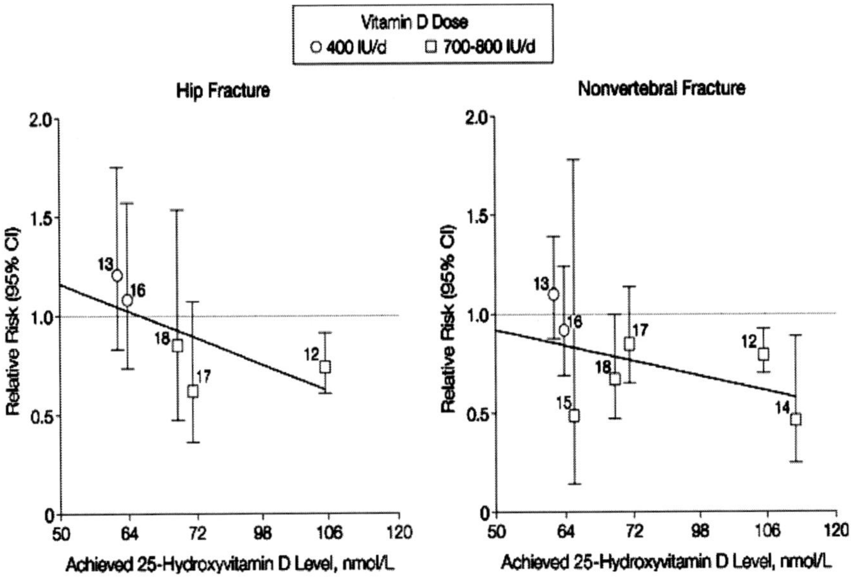

FIGURE 3-1 The effects of vitamin D supplementation on hip fracture and nonvertebral fracture.
NOTE: CI = Confidence Interval.
SOURCE: Bischoff-Ferrari et al. (2005). Copyright © (2005), American Medical Association. All rights reserved.

of hip fracture is observed in the area of 75–85 nmol/L (Figure 3-1). This raises the question as to whether it is possible to find biological markers or endpoints of this kind that will show a convergence of effects, where multiple goals of preventing fracture and perhaps contributing to the prevention of chronic disease can be embodied quantitatively in an endpoint.

Many avenues need to be pursued to better specify the selection of endpoints for reference values. This presentation has elaborated on some that may be useful and suggested that certain paths will be more fruitful than others. However, we must remember that one set of criteria or even an algorithm is unlikely to be "one size fits all" because there may need to be different approaches for different nutrients and types of reference values. This process will be an evolution that must be carefully planned.

General Discussion

A participant commented that the RDA and UL values are frequently close together because a UL is often established using an endpoint that occurs at a low level of intake for public health safety purposes, whereas endpoints selected for adequacy-based reference values tend to be those that occur at higher levels of intake. In response to the participant's question of whether this should be done in the future, Dr. Rosenberg replied that there should be even more collusion in the process of setting EARs/RDAs and ULs, especially because awareness of the margin between the two values is important. The decision should be driven by scientific data rather than the rote conclusion that the ULs must be as low as possible and the EARs/RDAs as high as possible.

An audience member remarked that it would be useful if the DRI study committees considered endpoints more comprehensively within their reports: For example, vitamin C at level X prevents scurvy and at level Y impacts another endpoint of interest. There would be different endpoints for the same nutrient, but endpoints more important in societies other than North America would not be neglected.

An audience member commented on Dr. Rosenberg's pessimism about using disease risk reduction for certain recommendations given the importance of reducing the risk of disease as an overall health goal. It was stated that there is a numerical relationship between fiber intake and the onset of cardiovascular disease (CVD). Another participant commented that measures of fiber intake from observational data can be a marker for other dietary and behavioral patterns and therefore may be problematic as a basis for setting DRIs. In response, Dr. Rosenberg noted that there will be instances when there is a direct relationship between dietary intake and a chronic disease response, but in many cases it will be difficult due to lack of specificity and confounding factors. He emphasized the importance of focusing on intermediate or surrogate markers predictive of disease outcome as a way of ensuring a focus on chronic disease risk reduction. A brief discussion took place regarding the process for validating biomarkers for disease. Dr. Rosenberg emphasized the need for sound science and clear validation.

DOSE–RESPONSE DATA: ARE THERE OPTIONS FOR DEALING WITH LIMITED DATA?

Presenter: Susan Taylor Mayne

A more challenging aspect of the DRI process is dealing with limited data on dose–response relationships. The DRI process depends on dose–response data for both EARs and ULs. Even if there are extremely limited data on dose–response for many nutrients, DRI study committees need to establish numeric values. As a consequence, some DRI values are "softer" in reality than what might be expected. This is well illustrated using the example of the dose–response data that were available in establishing the EAR for selenium.

Dose–Response Data and the Selenium Estimated Average Requirement (EAR)

The study committee considered several possible endpoints or biomarkers for selenium status, ranging from disease endpoints (e.g., Keshan disease and cancer) to blood or plasma selenium levels to plasma selenoprotein concentration as a biomarker of selenium status. The study committee ultimately chose plasma selenoprotein concentration maximization as the biomarker.

Two studies that evaluated maximization of plasma selenoproteins in response to supplemental selenium were available. One was a study of 52 men and women from New Zealand (Duffield et al., 1999), and the second was a study of 45 men from China (Yang et al., 1987). Both populations had low selenium intake.

In the New Zealand study, the baseline selenium intake of the subjects averaged 28 μg/day (for comparison, U.S. intakes are about 100 μg/day). Groups were given five different levels of selenium per day for 5 months: 0, 10, 20, 30, or 40 μg/day. The endpoint being monitored was plasma selenium-dependent glutathione peroxidase. All of the groups receiving additional selenium were found to have increased glutathione peroxidase, but they could not be distinguished from one another due to large variations in response. Because the variation was so large, a dose–response could not be calculated. Instead, the investigators decided that the lowest added intake, 10 μg/day, may be sufficient, so they set an EAR of 38 μg/day, which is the baseline intake of 28 μg/day plus 10 μg/day.

In the Chinese study, the baseline selenium intake of the subjects was even lower, 11 μg/day. Groups were given five different selenium doses for 8 months: 0, 10, 30, 60, or 90 μg/day. Although it was difficult to determine a dose–response based on the limited sample size, it was estimated that average maximization was achieved at the added intake of about added 30 μg/day. This gave an EAR of 41 μg/day when combined with the baseline intake of 11 μg/day. With weight adjustment to reflect North American body size, the EAR was increased to 52 μg/day.

The IOM study committee simply averaged these two numbers (38 and 52 μg/day), resulting in an EAR of 45 μg/day. As the variation data were difficult to calculate, a coefficient of variation of 10 percent was assumed, and the Recommended Dietary Allowance was set at 55 μg/day.

As discussed above, the EAR for selenium was based on fewer than 100 subjects. Dose–response data anywhere in the world were very limited. The only available data were obtained from selenium-deficient populations from outside North America. Important questions are: How relevant is this EAR to the United States and Canada? Are there alternative techniques that we should be employing to try to characterize dose–response using more relevant and statistically powerful data?

Solutions to the problem of limited dose–response data can be grouped into two general approaches. The first is the statistical or modeling approach, which applies various models to try to characterize dose–response, such as in relation to chronic disease or mortality (e.g., a large cancer prevention trial in the United States with 35,000 men randomized to selenium supplementation or a placebo). The second approach is the biological approach. Both approaches are described below.

The Statistical Approach

The advantage of the statistical approach is that many studies with large sample sizes are available (both observational and clinical trials). One disadvantage is that the intake data in these large population studies are often susceptible to measurement error. This is nutrient specific; for example, the intake data are not of good quality for vitamin E and selenium. However, in many of these same studies, we can examine plasma nutrient status as a biomarker for chronic disease risk to estimate the dose–response, which can then be related to intake data using metabolic or other relevant studies.

Different statistical approaches are used to analyze nutrients in relation to chronic disease risk. The traditional single-study approach is where one examines nutrient intake or status in relation to a chronic disease endpoint. The typical approach is to quantile the intake or status data, then examine the relationships across these quantiles and test for linear trends using statistical testing. Nutrient intake or status can also be examined as a continuous variable. The relationship between intake or status of nutrient X and disease Y can be modeled using regression. Both of these approaches typically assume a linear relationship, which may or may not be a valid assumption.

An example to highlight this is found in work from Ulrich (2007) relating folate status to breast cancer risk. Although some studies are finding protective effects with higher folate status, other studies are finding suggestions of adverse effects or at least no benefit. Ulrich (2007) has suggested this is because the relationship between folate and breast cancer risk is nonlinear (Figure 3-2). The linearity of a relationship depends on the part of the dose–response curve in which it lies (see dotted and dashed lines in Figure 3-2). This implies that one must be aware of the likelihood that many dose–response associations involving nutrients and chronic disease may be nonlinear.

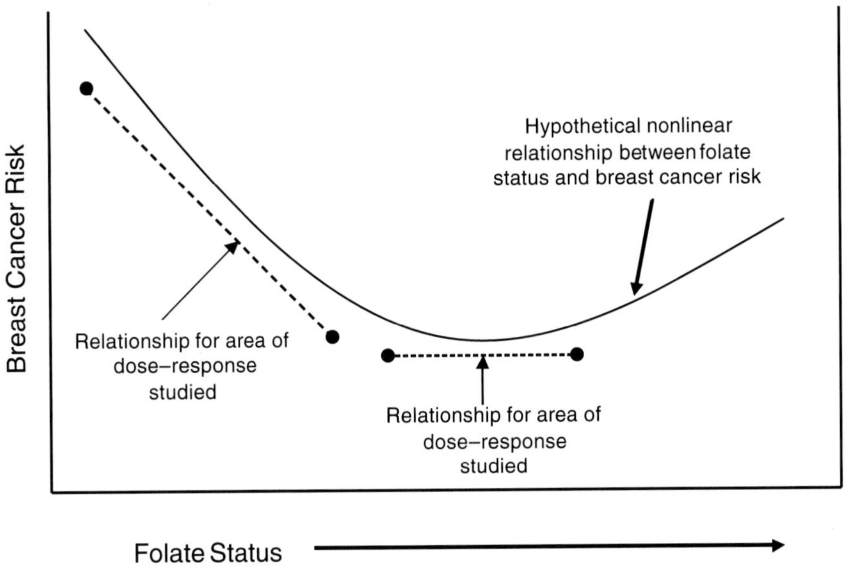

FIGURE 3-2 Hypothetical nonlinear relationship between folate status and breast cancer risk as compared with relationships for different areas of the dose–response curve.
SOURCE: Modified from Ulrich (2007).

One alternative to linear models is *restricted cubic spline models*, also known as *piecemeal polynomial curves*. Spline models allow for the examination of nonlinear effects of continuous variables (e.g., nutrient intake or concentration) in relation to disease risk. Some advantages of this approach are that no functional form needs to be specified; it is available in standard statistical packages (SAS, BMDP); and it can reveal nonlinear dose–response relationships.

An example of the use of restricted cubic spline models is from Wright et al. (2006), who examined the relationship between serum vitamin E and all-cause mortality (Figure 3-3). When the best model is fit to the data, as serum vitamin E concentrations rise, there is apparently a reduction in the risk of dying in this cohort up to a particular point; after that, it appears there is no additional benefit and, if anything, the possibility that the risk may start to increase. We might choose a serum vitamin E concentration associated with the minimum risk based on this curve, then determine the nutrient intakes required for half the population to achieve this plasma vitamin E concentration.

Combining data from multiple studies and using the data to estimate dose–response relationships are also possible. One standard approach is to

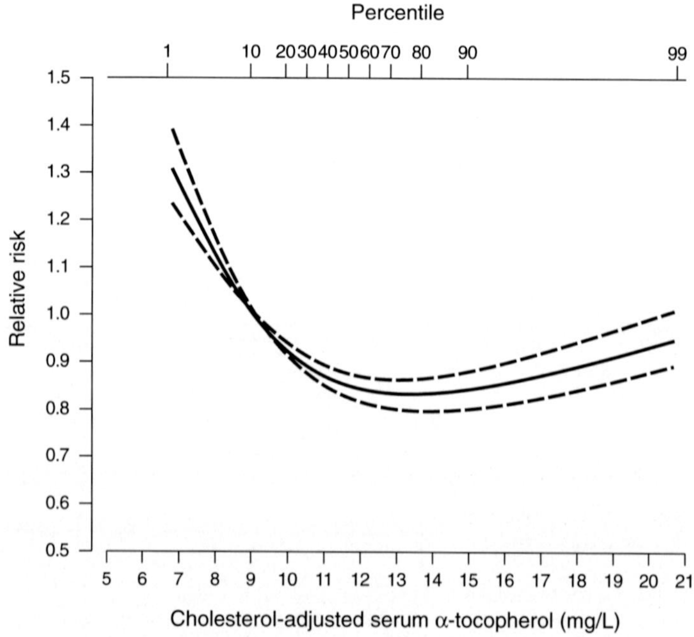

FIGURE 3-3 Cubic spline regression for total mortality according to cholesterol-adjusted serum α-tocopherol concentrations. —, Predicted relative risks; ---, 95% confidence interval. The reference value (9.1 mg/L; relative risk = 1.00) corresponds to the median value of the first quintile of serum α-tocopherol concentrations. To convert cholesterol-adjusted serum α-tocopherol concentrations from mg/L to μmol/L, multiply by 2.322.
SOURCE: Wright et al. (2006). *Am. J. Clin. Nutr.* (2006; 84; 1200–1207), American Society for Nutrition.

take data across multiple, randomized nutrient supplementation trials and perform a systematic review and meta-analysis. Meta-analysis was originally developed for clinical trials to see if an effect is present or not (e.g., do statins reduce CVD risk?). Meta-analysis can also be used to characterize dose–response using data from different trials with different nutrient doses and different achieved plasma concentrations.

In Figure 3-1 (page 67), from a meta-analysis looking at vitamin D supplementation and its effects on hip fracture and nonvertebral fracture, the authors performed a meta-regression to fit a linear regression to the data on the relative risk for a chronic disease endpoint as a function of achieved plasma 25-hydroxyvitamin D concentrations. Although they fit a linear model to these data, a nonlinear model may fit better, especially for hip fracture. We could have used these data to fit a nonlinear function, identify a plasma concentration at which lowest risk is observed, and then relate that level back to intake data.

Meta-analysis is also used for observational epidemiological studies of nutrients and chronic disease risk, but it was not designed for observational studies, and therefore its application is much more problematic. The dose that corresponds to high intake in one population may be very different from that in another population and in different parts of the dose–response curve (see Figure 3-2). The dose–response meta-analysis across categories can be done, with the caveat already mentioned. An example from the literature is a meta-analysis looking at observational studies on selenium intake and prostate cancer risk (Etminan et al., 2005). The investigators plotted studies of selenium intake (with lowest intake as the reference group) and risk of prostate cancer (Figure 3-4). Finding any dose–response data in this type of study is difficult because of the nonquantitative nature of the data.

Another approach to estimate dose–response is to combine data from multiple studies into a pooled analysis, where the original data from multiple studies are obtained and reanalyzed together. The assumption is that intake data across the studies are similarly (quantitatively) assessed, which is an assumption whose validity can be challenged. Validity is nutrient specific, depending on the ability to estimate intake of that nutrient accurately across populations.

An example of a pooled analysis is from Hunter et al. (1996), who examined the relationship between percentage of energy from fat in the diet and breast cancer risk (Figure 3-5) and concluded there was no association. However, it is assumed, perhaps not correctly, that when data are pooled from multiple cohort studies that use different dietary instruments, fat intake (along with energy intake) can be measured precisely and similarly across the studies.

The statistical approach can also apply to ULs. Instead of risk of inadequacy, risk of excess is modeled (e.g., the risk of hip fracture with high vitamin A intake). Similar approaches as described previously can be applied to ULs (e.g., spline models, meta-analysis, meta-regression), and the nutrient concentrations or intake levels at which risk of adverse effect begins to increase can be evaluated.

In terms of using chronic disease endpoints for dose–response estimation, although chronic disease data are widely available from U.S. and Canadian populations, causality and confounding (e.g., correlated nutrients from the same foods) are difficult to address. The use of plasma biomarkers is desirable to examine dose–response, but it does not solve the confounding problem.

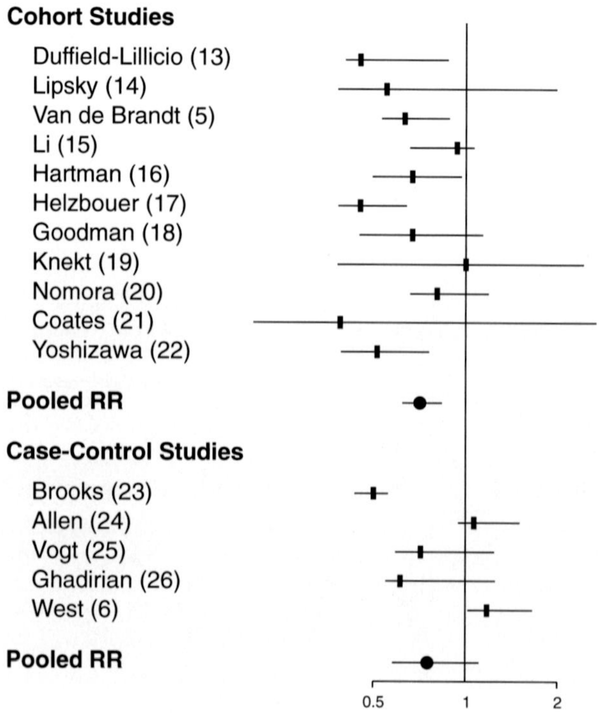

FIGURE 3-4 A meta-analysis of observational studies of selenium intake and prostate cancer risk.
NOTE: RR = Relative Risks.
SOURCE: Etminan et al. (2005). Reprinted from *Cancer Causes and Control* 16:1125–1131, figure 1, with kind permission from Springer Science and Business Media. Copyright © Springer 2005.

The Biological Approach

The biological approach uses the mode of action framework. The idea is that in order to approximate a dose–response, we need to understand the mode of action of nutrients. This has a straightforward application to ULs, but it can apply equally to nutrient deficiency. Key molecular and biological systems and pathways that are modulated by nutrients need to be identified.

The background paper on the biological approach ("Approximating Dose–Response in the Face of Limited Data," posted on the IOM website [www.iom.edu/driworkshop2007]) describes the tools and technologies that are in use in other fields that may be helpful in establishing dose–response

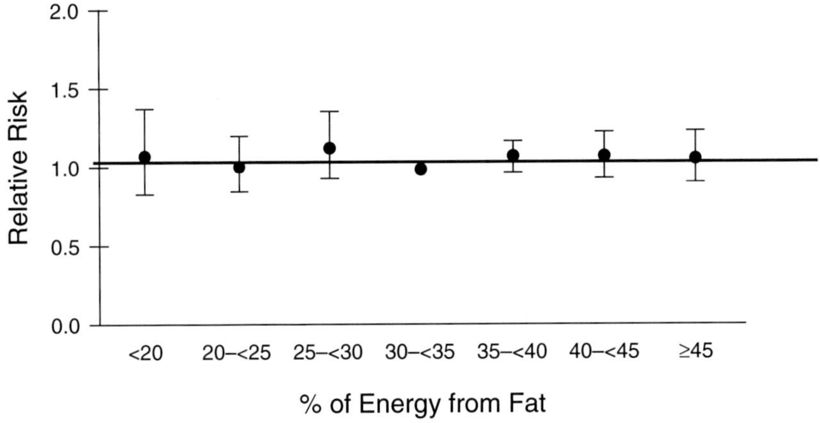

FIGURE 3-5 Fat and breast cancer: pooled analysis.
SOURCE: Hunter et al. (1996).

in the nutrition literature. Mapping pathways could be helpful. However, we often know the pathways that cause deficiency, so mapping pathways may not necessarily get us closer to dose–response. In vitro tests in human cell lines are being widely used in the pharmaceutical industry, for example. However, human dose data supersede in vitro tests in human cell lines. High-throughput methods are another technique where, for nutrients, different doses could be used to see if any inference or input about dose–response can be obtained, but they will not necessarily solve any problems. Microarrays, computational biology, and physiologically based pharmacokinetic and pharmacodynamic models are all reasonable approaches where animal models of nutrient toxicity and deficiency are available and are particularly helpful for different life stages. Not mentioned in the background paper are metabolomics and translational biology, fields where there is much research progress that may be helpful in terms of informing us about dose–response in the future.

Conclusion

There is a real trade-off between the statistical and biological approaches. The statistical approach targets the right population and the right nutrients, but there is limited causal inference when dealing with chronic disease endpoints. The biological approach is mechanistically driven, but there is a tenuous link to the human dose data.

In conclusion, newer options exist for examining dose–response in the setting of DRIs, but none are yet mature or ideal. To move forward, we must have a multidisciplinary, integrated approach involving biostatisticians, toxicologists, and nutrition scientists. There is no obvious advantage of one approach over another at this time. An approach of data convergence, where we look at all of the evidence to determine if we can characterize dose–response, may be most useful.

General Discussion

One audience member raised the issue of meta-analysis studies and sensitivity testing, and asked whether Dr. Mayne saw any role for sensitivity testing and, if so, what sort of new data would be needed to determine the quality of the current body of data or to change overall findings. Dr. Mayne responded that sensitivity analysis usually examines whether a particular study has undue influence on the results and whether its inclusion or exclusion changes conclusions about the existence of an effect. Regarding her focus on dose–response data, Dr. Mayne indicated that it would not necessarily be informative to do a meta-analysis based on excluding or including specific trials to examine dose–response relationships among the studies. The audience member further speculated on ways to determine if small studies or a large study is needed.

In response to a request for clarification on the New Zealand and Chinese selenium studies, Dr. Mayne said she was not familiar with all of the details of the studies. While both studies were published, they were not in a readily accessible journal at the time the committees initiated their work. She noted that the Chinese data were published in a book, and the New Zealand data were analyzed for the DRI report and subsequently published in 1999. She commented that nutrient deficiency data tend to be limited, often from other countries, and frequently outdated. However, she noted in contrast that an intervention study on selenium involving 35,000 people is currently underway; it is a 13-year clinical trial that completed recruitment about 2 years ago.

Another question was related to the earlier comments about the challenges associated with chronic disease endpoints for DRIs because the available studies often lack dose–response data. The question was raised, given Dr. Mayne's presentation on tools to approximate dose-response relationships, as to whether it was premature to suggest eliminating chronic disease endpoints because they lack dose–response data. Dr. Mayne responded that some tools are available for this purpose, but the confounding issue remains for certain chronic diseases. For example, nutrients are only one of numerous factors that determine cancer risk. However, for some chronic diseases there may be a nutritional role more proximal to the disease endpoint, such as lutein and zeaxanthin for macular degeneration. Therefore it is unwise to "throw out" all chronic diseases because some are multifactorial and confounded. In short, chronic diseases should not be lumped together in considering their utility in the DRI process; as endpoints they likely need individual consideration.

WHAT ARE THE CHALLENGES IN ADDRESSING EXTRAPOLATION/INTERPOLATION FOR UNSTUDIED GROUPS?

Presenter: Stephanie A. Atkinson

The paucity of data for certain subpopulations resulted in extensive use of extrapolation models during the DRI development process. In fact, about 60 percent of the DRIs were derived by extrapolation for 1- to 18-year-olds. The paucity of specific data available based on research in infants and children is concerning. For this reason, careful consideration of extrapolation methods is needed to ensure that we are doing the best we can until that point when data are available and DRI reference values can be set without the need for extrapolation.

Our experience suggests that various approaches to extrapolation have been used, which has led to inconsistencies in reference values among age groups. For example, for the 6- to 12-month age group, extrapolation up from the AI for 0- to 6-month-olds was done for niacin, choline, biotin, and vitamins B_{12}, A, and K. At the same time, extrapolation down from the adult EAR or AI was done for vitamins B_1, B_2, and B_6; folate; and pantothenic acid.

Furthermore, extrapolating down from adults, with inappropriate models in particular, leads to DRIs that do not make much sense. For vitamin A, the AI is 500 μg retinol activity equivalent (RAE)/day for 6- to 12-month-olds compared with an RDA of 300 μg RAE/day for 1- to 3-year-olds, one being extrapolated up from the 0- to 6-month group and the other being extrapolated down from adults. For vitamin C, the AI is 50 mg/day for the 6- to 12-month group based on the composition of human milk and intake from food, but the RDA is only 15 mg/day for the next age group, being extrapolated down from adults. An effect in the opposite direction is observed for the derivation of DRIs for molybdenum: an AI of 3 μg/day for 6- to 12-month-olds based on human milk and food and an RDA of 17 μg/day for 1- to 3-year-olds extrapolated down from adults.

In the case of fiber, there were no data for young children, so the AI was extrapolated down from an adult AI of 14 g/1,000 kcal based on the reduction in CVD risk. The AIs for children 1–13 years of age range from 19 to 31 g/day, whereas the intakes obtained from diet surveys range from 5 to 18 g/day (Suitor and Gleason, 2002; Devaney et al., 2004). Clearly young children and adolescents are not consuming anywhere near the amount of fiber predicted by extrapolating down from adults. Thus, the AI may be impossible to achieve; more importantly, it may not be physiologically appropriate.

Overview of Available Approaches

In North America, *extrapolation* is used in DRIs to adjust for physiological differences between groups of varying body size or age to establish a reference value for an unstudied age/gender group. In Europe, *scaling* has been used since the early 1800s with regard to expressions of body weight or compartments scaled to height (e.g., the body mass index, or BMI). "Scaling" may be the more appropriate term to use for the purposes of DRI development.

Regardless of terminology, the types of extrapolation/scaling models used are fairly similar. There are linear models, where body size (mass) can be used with a reference body weight for age and gender (e.g., the AI for fiber for children) or as a function of energy, where the median reference energy intake for age and gender is used (e.g., the AI for water, sodium, and potassium for children).

The problem with the linear model is that there is no accounting for age variations in intermediary metabolic rates, energy intake, or basal metabolic rate (BMR). The exponential model, on the other hand, tries to adjust for metabolic differences related to body weight (BW) (e.g., the UL extrapolation from adults to children uses $BW^{0.75}$). The issue is that this assumes that maintenance needs of nutrients as a function of metabolic weight are similar for adults and children and similar across genders. It also assumes that absorption, digestion, and excretion are similar across age groups. Apparently there is a lack of consensus on which adjustment factor best reflects BMR (e.g., adjustment factors in the range 0.6–0.8 have been used). The values produced by the exponential model are always higher than those produced by the linear model.

The other model for scaling is relative to body surface area, and this adjusts for metabolic differences between ages related to body surface area based on its relation to BMR. This will always result in higher nutrient reference values than those based on body weight. A study by Przyrembel (2006) shows a nearly twofold difference between relative nutrient intakes using body weight and those derived using body surface area for children up to 1 year of age (Table 3-1).

Some reports, such as that of the Scientific Committee for Food of the European Union (SCF, 1993), use *interpolation*, which is different from scaling or extrapolation, in that the value is interpolated for an age group between known values of age groups older and younger. Which of these models is most accurate is open to interpretation.

In all of these models, few other factors can be applied differently across reference intake standards of various agencies. One of those is the values used for growth in extrapolating from adults to children. In the DRI reports, we used approximate proportional increase in protein requirements for growth as established by the World Health Organization (WHO, 1985); the growth factors were 0.30 for 7 months to 3 years and 0.15 for older age groups. These were applied for every nutrient by assuming that the growth factor was the same as that established for protein.

The other variable is reference body weights, which should reflect those of the country or countries to which they are being applied. The reference body weights changed during the DRI process. At first, National Health and Nutrition Examination Survey (NHANES) III values were used (IOM, 1997); later, when the new Centers for Disease Control and Prevention (CDC)/National Center for Health Statistics (Kuczmarski et al., 2000) values became available, median heights and weights computed from median BMI were used. Infants were not based on the breastfed infant population because the data were not available (the CDC data mostly reflect formula-fed infants). For the next round of DRIs, we could use the recently published WHO reference growth standards for breastfed infants (Garza, 2006; de Onis et al., 2006, 2007) for 0- to 5-year-olds or perhaps Canadian reference growth data, which may be available soon.

Challenges

Extrapolation is a proxy; thus, it yields a risk of error, especially when values are extrapolated from adults to children. The evaluation of physiology requires scientific judgment. We need knowledge of substrate absorption, metabolism, and deposition in tissues during growth phases and renal and other excretion that may affect the EAR or UL, as these may not be related in a simple fashion to body size, even in the exponential model. For some nutrients, especially for the UL, extrapolation on the basis of body weight or body surface area yields a UL for children that is incompatible with known nutrient intakes. Perhaps ULs should not be set for children until we have direct experimental evidence.

Children are not just little adults. They need their own evidence-based DRIs. If DRIs are inappropriately set, we might adversely affect the health of children. We could identify the wrong nutrient intake problems (either inadequacy or excess), which could lead to inappropriate recommendations for child health feeding programs (e.g., Special Supplementation Nutrition Program for Women, Infants and Children) and have a public health impact.

TABLE 3-1 Relative Nutrient Intake (RNI) Reference Values by Extrapolation from Adults: Body Weight Versus Surface Area

Age	RNI Based on Child/ Adult Body Weight	RNI Based on Child/ Adult Body Surface Area
Newborn	0.05	0.11
0.5 years	0.10	0.19
1 year	0.14	0.23
10 years	0.46	0.59

SOURCE: Przyrembel (2006).

Conclusion

In an ideal world, the use of scaling and extrapolation models in setting DRIs should be unnecessary. However, the reality is that they must be used. In such cases, we need to ensure that we are using biologically plausible models and recognize the role of well-reasoned and transparent scientific judgment.

Today there are opportunities that did not exist 15 years ago to conduct research in children, including the use of stable isotopes to measure energy requirements, amino acid oxidation, and amino acid requirements, as well as for trace element turnover. The pursuit of these appropriately designed studies is critical.

General Discussion

One person pointed out that obtaining data for currently unstudied groups will take a long time, even with new methods. Given that, he asked how we can use data from animal models, which are more readily available and can be obtained in a shorter time. Dr. Atkinson responded that animal models can be helpful, but need to be closely aligned to the human infant or young child, such as monkeys or piglets. These are expensive research models. An audience member then questioned whether, given the difficulties and lack of data, ULs for children should not be developed. Dr. Atkinson suggested it would be important to pursue appropriate animal models to study adverse effects as a preliminary and hypothesis-generating step for this purpose. The audience member suggested that at least two or three species should be used to reduce uncertainty.

A participant remarked that stable isotopes are an excellent approach for children. However, noting they are also expensive, he asked whether marker nutrients might be translated to other nutrients in the same category to lower costs. Dr. Atkinson agreed it was possible. She also noted that a practical barrier in doing research in normal children is the ability to draw blood, and that obtaining urine is less challenging.

The comment was made that Dr. Atkinson did not address pregnancy and lactation as extrapolation concerns in her presentation. She responded that extrapolation for pregnancy and lactation requires a different focus from her main topics for this presentation. She noted a reference that would be helpful regarding pregnancy and lactation (Atkinson and Koletzko, 2007).

WHAT ARE THE CHALLENGES IN ADDRESSING ADJUSTMENT FOR DATA UNCERTAINTY?

Presenter: Hildegard Przyrembel

The root of the problem we face is a lack of data, which results in uncertainty. Uncertainty can be reduced only by the acquisition of more data.[2] Alternatively, analysis of the impact of sources of uncertainty can be used to understand and thereby help to address uncertainty. The analysis can be qualitative (descriptive) or, preferably, quantitative (mathematical modeling). This presentation will focus on adjusting for data uncertainty from the perspective of establishing ULs, but many of the principles may also apply to establishing reference values.

For the purposes of establishing ULs, the mode of action (i.e., endpoint of interest) and the related dose–response relationship are critical pieces of information, as shown in Figure 3-6. A no-observed-adverse-effect level (NOAEL) or lowest-observed-adverse-effect level (LOAEL) is identified (or, alternatively, a benchmark dose is calculated), then modified by the use of uncertainty or adjustment factors in order to derive the UL. *Uncertainty factors* refer to default values with no or little factual basis, whereas *adjustment factors* are values supported by actual toxicodynamic and/or toxicokinetic data.

On the other hand, establishing requirements and avoiding deficiencies also require an understanding of the mode or mechanism of action as well as information on the dose–response relationship (see Figure 3-7). However, the identification of the mirror image of the NOAEL or LOAEL (i.e., a critical dose) would be problematic. Assuming it could be determined, then it would need to be multiplied by an adjustment or uncertainty factor in order to obtain the "lowest threshold value of intake." A question would remain, however, as to how to convert this value into an average estimated requirement, the most obvious but perhaps unsuitable suggestion being that

[2]*Data uncertainty* must be differentiated from *data variability*, which is due to the heterogeneity of a quantity over time, space, or members of a population. It can be reduced by selection of the sample, not by the provision of more data. However, probabilistic assessment methodology is now used widely for assessing data variability.

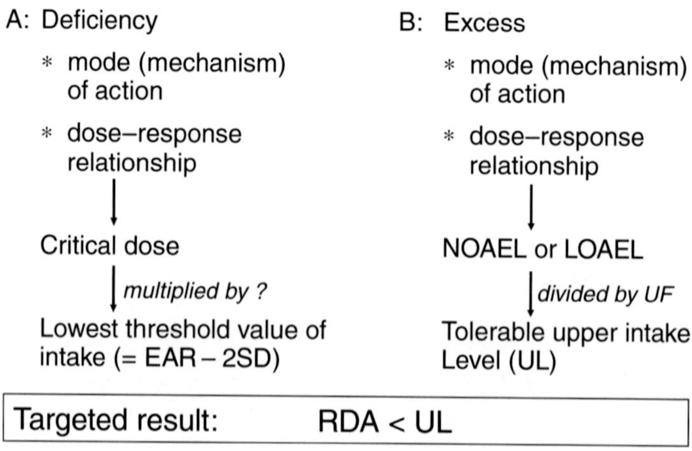

FIGURE 3-6 Risk assessment of essential nutrient and adverse health effect.
NOTE: EAR = Estimated Average Requirement; SD = standard deviation; NOAEL = no-observed-adverse-effect level; LOAEL = lowest-observed-adverse-effect level; UF = uncertainty factor; RDA = Recommended Dietary Allowance.

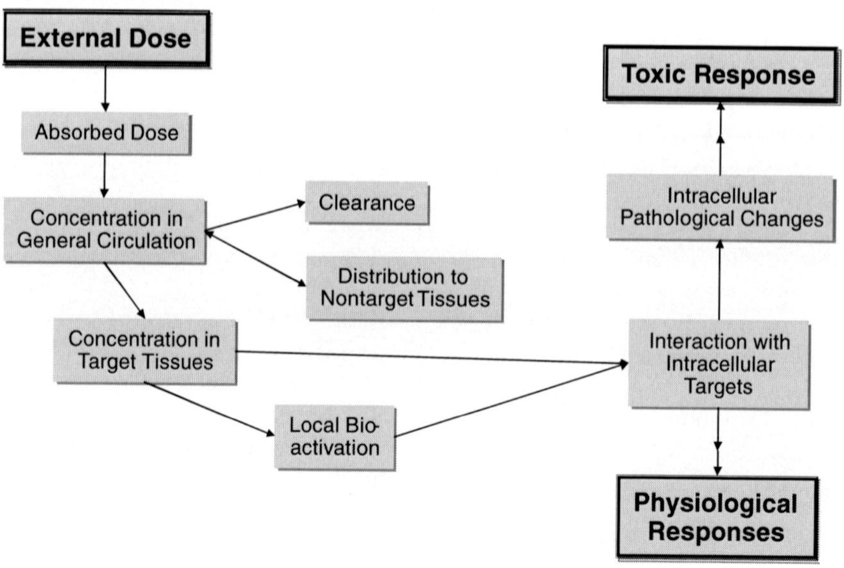

FIGURE 3-7 The multiple steps between intake of a nutrient and either physiological or toxic responses, depending on dose.
SOURCE: Modified from IPCS (2005).

this lowest threshold value reflects a value two standard deviations below the median requirement.

Considering the Points of Impact for Data Uncertainty

In general, from the metabolic and physiological perspectives, the points at which uncertainty can have an impact and therefore be assessed are divided into multiple steps between intake of a nutrient and either physiological or toxic responses, depending on the dose (Figure 3-7). Most of the literature and research on uncertainty analysis have been done for chemicals, rather than nutrients. This rendition is derived from those fields of study. The pathway moves from the "effect" of the dose and separates at the interaction of the nutrient or metabolite of that nutrient with intracellular targets, to go toward either the physiological response or a pathological or toxic response. All the different steps can be characterized either in animals or in different age or gender groups of humans, which helps to modify or quantify the necessary adjustment or uncertainty factors. However, few data are available on these different steps, making it difficult to develop reasonable adjustment factors.

Another illustrative example from the field of chemical study that is inexplicably missing from work in the nutrition area is a theoretical dose–response curve for various effects occurring in the population. This type of mapping greatly assists efforts to study and address uncertainty. For instance, Figure 3-8 plots the percentage of the population with an effect against the range of acceptable daily oral intake of a nutrient; it shows that different endpoints can be identified, and it suggests that these dose–response curves should be parallel (although there is no reason why they should be parallel).

Further, the same stepwise procedure for increase of effects could apply for both nutrient intakes higher than the requirement (excessive) and intakes lower than the requirement (deficient). Figure 3-9 shows a combined curve of dose–response relationship for the risks due to deficiency (absence of benefit) and toxicity. What is a benefit? There is no assurance that a benefit will result from higher intakes or that the benefit always needs higher intakes and requirements. What is needed for this kind of parallel assessment of excess and benefit is the intake that gives a 50 percent incidence response, the ED50, and the coefficient of variation (CV) of response. In the graph, different CVs have been assumed (as they often are). The CV influences the point where the two curves for toxic response and benefit response intersect.

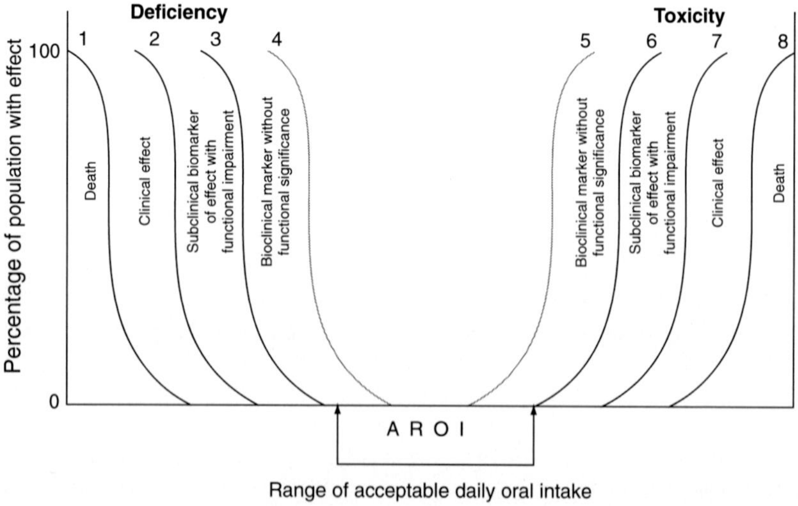

FIGURE 3-8 Theoretical dose–response curves for various effects occurring in a population.
NOTE: AROI = Acceptable Range of Oral Intake.
SOURCE: IPCS (2002).

Sources of Uncertainty

The main sources of uncertainty include the model used, the quality of the available data, and the scaling algorithm. The model used is determined by both structure and parameters. Sensitivity analysis can be performed to identify those parameters with significant impact on model output. Regarding available (or input) data, they are often lacking and need to be evaluated in terms of their quality, variability, and measurement and database errors. Quantification of missing data is impossible. The only solution is to input fictive or virtual data into the models and assess the difference in outcome. For scaling algorithms, it is always difficult to be certain that the algorithm chosen is appropriate. One problem of scaling is that it propagates errors made earlier in the process.

Turning more specifically to the DRI development process, uncertainties in available data include methodology of balance studies (e.g., calcium), lack of data (e.g., pantothenic acid), physiological significance (e.g., vitamin K), and lack of identification of an adverse effect (e.g., vitamin B_1). Major sources and types of uncertainties in dietary exposure assessment include food consumption, body weight, and content in food. Uncertainty in relation to the food composition database is large and relates to fac-

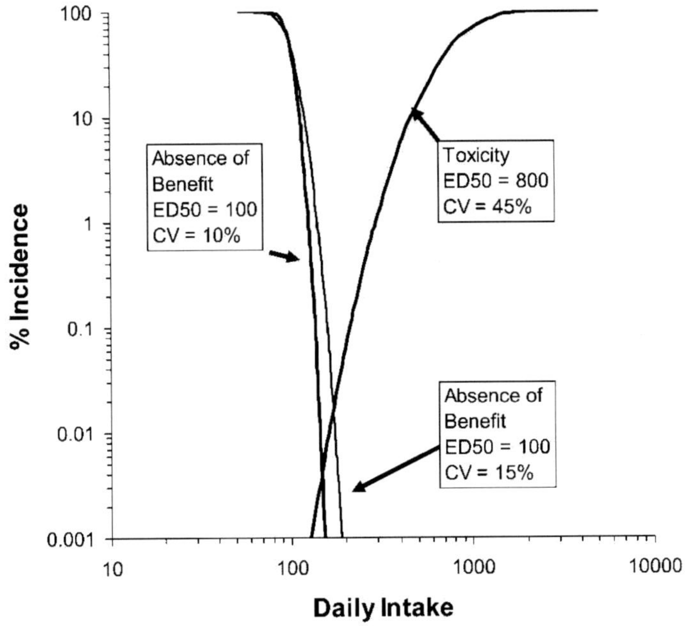

FIGURE 3-9 Dose–response relationships for the risks due to the absence of benefit or the presence of toxicity. The data were plotted assuming a log-normal distribution. The absence of benefit (equivalent to a deficiency condition) has been plotted assuming coefficients of variation of 10% (thick line) and 15% (narrow line), and the toxicity line has been plotted assuming a coefficient of variation of 45%. The intersection of the two lines is the optimum intake, provided that the nature of the deficiency and the toxicity are of equivalent adversities. ED50, the dose that gives a 50% incidence; CV, coefficient of variation.
SOURCE: Renwick (2006). *J. Nutr.* (2006; 136; 4935–5015), American Society for Nutrition.

tors such as differences in bioavailability of nutrients from different foods and variability in composition with, for example, storage, processing, and preparation.

The uncertainty factors that have been used traditionally in chemical toxicology are default values—for example, 10 for extrapolation from animals to humans (interspecies) and 10 for coverage of human variability (interindividual). There are other uncertainty factors for use of subchronic rather than chronic studies, use of a LOAEL instead of a NOAEL, and deficiencies in the database. The applicability of these to nutrient considerations is questionable (especially given the dual risk of essential nutrients—risk from too little and risk from too much), and they need considerable modification for use with nutrients, assuming they are useful at all.

In the DRI reports, where the endpoint was a human NOAEL, an adjustment factor of 1.0 was chosen in half the cases. The justification for using an uncertainty factor above 1.0 was mostly variability in the population or insufficient data. Where a human LOAEL was used, the uncertainty factors were higher, except for magnesium, fluoride, and sodium, and the justification for the larger uncertainty factor was often the use of the LOAEL. Where animal LOAELs were used, the variation in the selection of adjustment factors was especially great.

Conclusion

Uncertainty analysis can help those responsible for developing DRIs. It is intended to systematically examine the adequacy of the selected model (e.g., to see if predictions agree with observations), the uncertainties in model parameters and input data (using mathematical methodologies), and the presentation of the results (e.g., probability distribution). Communication of the uncertainties and how they have been compensated for are very important. Of course, uncertainty analysis does not preclude the need for appropriate exposure data, relevant endpoints, and trustworthy dose–response data.

In conclusion, uncertainty in nutrient risk assessment and in the definition of requirements of nutrients is unavoidable. It should be characterized with respect to its nature and magnitude, and different types of uncertainty should be ranked according to their impact on the results of the procedure. For nutrients, the default uncertainty or adjustment factors conventionally used in the risk assessment of chemicals have to be modified, ideally by either chemical-specific kinetic or process-specific dynamic data. We should try to obtain these data, but we will have to do this for each nutrient individually because of the different and multiple physiological functions of different nutrients in the human body. Uncertainty due to variability in both kinetics and dynamics can be dealt with by mathematical procedures. Uncertainty due to gaps in data, however, can be effectively relieved only by the acquisition of more data.

In the meantime, assumptions about data used in the assessment and the impact of their intentional variation in the calculation need to be identified and communicated both qualitatively and quantitatively. However, how this communication is understood by the users is uncertain.

General Discussion

A participant noted that one of the slides showed that the CV for benefit was assumed to be either 10 or 15 percent, whereas the CV for toxicity was apparently assumed to be 45 percent. He expressed the view that none of the ULs are set on the basis of a measure of central tendency; instead, the ULs are based more on a threshold concept. He queried whether Dr. Przyrembel concurred with the validity of the graph. Dr. Przyrembel responded that the author of the slide justified his selection in his publication, giving data to support his assumption that variability in sensitivity to toxic effects differs from variability in requirement. The participant then outlined a different approach (when there are sufficient data) based on rank ordering the clinical trials and omitting the use of numerical adjustments.

An audience member asked how to deal with the uncertainty associated with studies that administer similar doses, but demonstrate different responses. Dr. Przyrembel responded that the only solution is to carefully examine the study design for an explanation of the differences.

Another participant raised a question about using clinical studies that were designed for efficacy or benefit trials to ascertain adverse event information. Evaluating the equivalency of studies is very difficult if they have been conducted for a different purpose.

ESTIMATING DIETARY INTAKE: WHAT ARE THE IMPLICATIONS FOR DRI DEVELOPMENT?

Presenter: Amy Subar

This presentation addresses the use of dietary intake estimates in DRI development, notably as it relates to the step focused on dietary exposure (or intake) assessment. These estimates of current intake in the United States and Canada allow study committees to place DRI values once they are developed within the context of the population's current estimated consumption and, in turn, to characterize the risk of inadequate or excessive intake. An understanding of the strengths and weaknesses of the various dietary assessment methods for estimating current population intakes is important in ensuring the proper use and interpretation of these dietary estimations.

Other types of data on intake relevant to DRI development include studying dose–response relationships in clinical feeding studies, evaluating DRIs in population-based epidemiological or clinical studies, and developing AIs from national dietary surveys, such as NHANES. These types of data and the use of estimated intakes to examine the relationship between intake and health outcome will not be specifically discussed.

Methods for Estimating Intake: Self-Report Instruments

It should be noted that the goal for all applications of dietary intake estimation is an estimate of usual intake, which is the theoretical long-run average daily intake of a dietary component. Three main types of self-report instruments are used to collect such data: 24-hour recalls, food diaries or records, and food frequency questionnaires (FFQs).

Twenty-Four-Hour Recalls

Twenty-four-hour recalls can vary in many ways. Training of the interviewers and standardization of probing questions (i.e., questions that follow after someone reports eating a particular food, such as what kinds of fats were added to foods) can vary from study to study. Most 24-hour recalls are collected by some sort of standardized computerized approach, but some studies use pencil-and-paper administration with later coding of the data. Some recalls are done in person, others by telephone. Different kinds of portion size models or measurement aids are used to estimate portion size.

The 24-hour recall has various strengths. The intake data can be quantified in detail. In theory, it should not affect human eating behavior because the respondents are asked to report what they ate yesterday, intake that would have occurred before they knew they would have to report such intake. There is lower sample selection bias than for other methods because the recall does not require literacy and the respondent burden is low. It is generally agreed that this is the most reliable method for dietary assessment. Furthermore, usual intake distributions can be estimated from as few as two dietary recalls.

One weakness is that recalls rely on memory. Also, 24-hour recalls are costly to develop and administer because highly trained interviewers are needed. In addition, because recipes and preparation methods vary for many foods, default recipes and hence nutrient values are used, and these may not accurately capture the level of nutrients consumed. Underreporting of foods and amounts eaten is also common, especially among those who are overweight or obese. Finally, at least 2 days and statistical modeling are required to obtain usual intake estimates.

Food Diaries or Records

Food diaries or records are, in general, less standardized than dietary recalls. Respondents do not have to be trained, but the diaries may or may not obtain comprehensive data, and the coding of those data is highly variable from study to study. The use of technology to collect real-time dietary data has been a research topic of great interest, with technology such as personal digital assistants, cell phones to take pictures, and voice recognition being explored.

If done correctly, a food diary or record can provide quantified and detailed intake information. It can be relatively accurate, and it is done in real time so in theory should not rely on memory. The biggest weakness of a food record is that it is reactive and hence biased. Because respondents know they have to record, they may change what they eat because it is difficult to record, or they may undereat. The food record requires literacy, and it has a high respondent and investigator burden. There is a high sample selection bias because only certain people are willing to keep records. The longer people keep records, the worse the data quality is. Although it should be real time, people often record the data at the end of the day. Underreporting is typical, and worse with those who are overweight or obese.

Food Frequency Questionnaires

In the often self-administered FFQs, people are asked a series of questions—usually hundreds—about how often they usually consumed a particular food in a given time period; what preparation methods were used; and what the typical portion size was. These components vary among FFQs, as do procedures to determine the food list and the nutrient composition assigned to each food. One strength of the FFQ is that the respondent burden is relatively low because the questionnaire is filled out only once. The focus is generally usual intake and the total diet. An FFQ should not be biased by changes in eating behavior because intake in the past is queried. Another benefit of the FFQ is the low cost associated with administering the instrument and processing the data.

One weakness of the FFQ is that it lacks detail because it contains a finite list of foods and details are not generally collected. It is cognitively complex for respondents to report what they ate over the past year, for example. It requires literacy. Different FFQs can produce different results in the same population, whereas the same FFQ can produce different results in different populations. There is severe measurement error when looking at absolute intakes. To reduce this bias, epidemiologists rank individuals and adjust the models for energy intake. In general, outcome findings are attenuated by the amount of error in the FFQs.

Methods for Estimating Intake: Biomarkers

Certain so-called "biomarkers of intake" may be used to assess dietary intake. A *recovery biomarker* is one in which there is a 1:1 relationship between what is consumed and the biomarker value. Such biomarkers provide very accurate data on what individuals are consuming, but few of these can be used: doubly labeled water, urinary nitrogen, and possibly urinary potassium.

Concentration biomarkers reflect a direct biological response to what someone consumes. It is more of a correlated response, and it is affected by other characteristics of the individuals (e.g., whether they smoke or possibly their body weight). Therefore, it cannot be used to assess the amount consumed, and it may reflect short- or long-term intakes. In general, it is difficult to use such biomarkers to evaluate direct dietary intake for purposes of DRIs.

There are also *homeostatically controlled biomarkers*, which have no direct relationship to intake.

Challenges and Sources of Error/Bias

First, underreporting occurs in all of the self-report dietary assessment methods described above. The percentage of energy underreported based on a review of doubly labeled water studies was up to 58 percent for food records, 38 percent for FFQs, and 26 percent for 24-hour recalls (Trabulsi and Schoeller, 2001). Underreporting can vary by gender, age, and BMI. In general, underreporting tends to increase as body weight increases. For example, results from NCI's Observing Protein and Energy Nutrition (OPEN) study, conducted with about 500 men and women using doubly labeled water and urinary nitrogen, show that energy underreporting occurs for both 24-hour recall and FFQ, and is greater for FFQ (Figure 3-10). The results also show that underreporting varies by BMI for the FFQ and 24-hour recall (not shown).

Second, data on dietary supplements may not be collected in many studies. We have to assume that measurement error is present in assessing self-reported dietary supplement intake. However, not accounting for supplement intake leads to substantial underestimation of total nutrient intake. When supplement intake is included, this results in highly skewed intake distributions, which present challenges for describing usual intake distributions.

Another source of error in all self-report dietary data relates to the nutrient database. Analytical methods for nutrient composition change and improve, and, just as importantly, the composition of finished food products is constantly changing. Therefore, the database that we use needs to be updated and to match the time period of the study.

Obviously it is impossible to observe long-term or usual intakes. Rather, the approach is to acquire estimates based on statistical modeling using

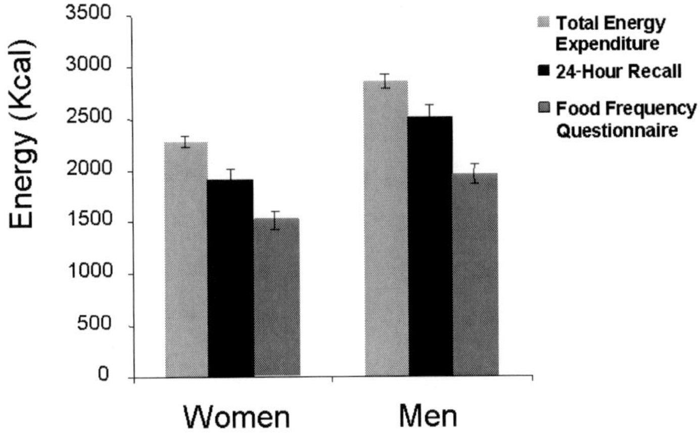

FIGURE 3-10 Results from the Observing Protein and Energy Nutrition (OPEN) Study: Energy intake underestimation by 24-hour recall and food frequency questionnaire compared with total energy expenditure.
SOURCE: Subar et al. (2003).

short-term, self-reported data. Early in the evolution of dietary intake estimation, we used a single day of intake and called it usual intake based on recalls from national surveillance studies. Then we realized we needed at least the average of a few single-day measurements to improve estimates. Next, we became more sophisticated and used statistical modeling: first the NRC method, then the Iowa State University (ISU) method, and more recently the NCI method.

Given that the assumptions involved are taken into account, these statistical models remove day-to-day variability from the 24-hour recall so that a better estimate of usual intake is obtained. This is illustrated in Figure 3-11. The probability is plotted against the usual intake of energy; 2,200 calories is the cutpoint. If 1 day of intake is used, the distribution would be long and skewed to the right side. When statistical modeling is applied—removing some of the variability—a more normal distribution of intake is obtained, as would be expected in the population as a whole. This statistical treatment of the data is important, and methods continue to be developed to establish usual intake distributions. The NCI method builds on the NRC/ISU methods to estimate usual nutrient intake distributions. It can also handle episodically consumed dietary constituents, such as vitamin A, and it can be applied to foods and dietary supplements. It also provides greater power to conduct subgroup analyses within the same model.

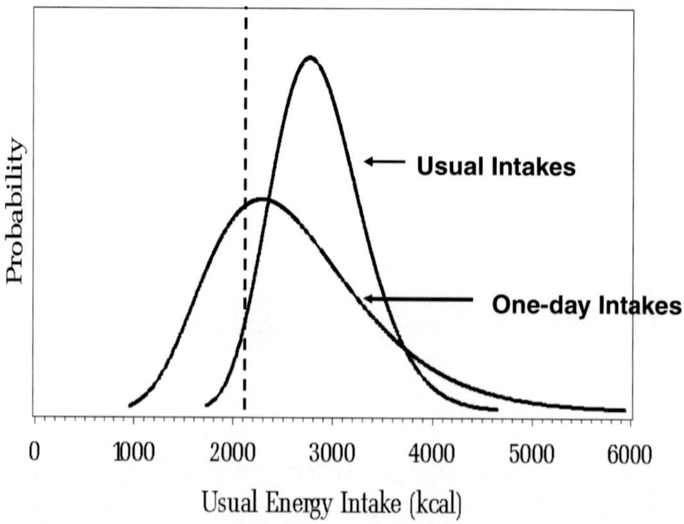

FIGURE 3-11 Probability of consuming above or below cutpoint (dashed line): One-day versus usual intake distributions.

Implications

Dietary exposure (or intake) assessment for DRI development ideally would be based on usual intake distributions estimated from some multiple days of intake and statistical modeling. Sometimes there is interest in using intake data from observational studies. However, we have to be careful, given the amount of error that can occur in FFQs and other methods used in such studies.

The starting point for DRIs is the available clinical and metabolic data concerning requirements, health outcomes, and adverse events; DRIs are not derived (AIs excepted) from estimates of usual intake. Therefore, it is understandable that DRIs, even when developed using the best available scientific data, might be disparate from estimated intakes from dietary surveillance data. A clear understanding of the strengths and limitations of the dietary intake estimates allows those responsible for DRI development to put the scientifically derived DRI values in the context of current estimated intakes and, in turn, advise users of DRI values about differences between values and estimated intakes and possible reasons for them; it also identifies avenues for further research.

General Discussion

An audience member questioned whether progress can be made as long as we rely on people to report dietary intake information. Dr. Subar emphasized that the data are not all poor and that newer advances have shown considerable promise for ensuring good-quality estimates of intake. She pointed out that even though there is some level of underreporting, better ways to adjust the data are likely to be developed. The key point is that existing data need to be used appropriately, with an understanding of their limitations. Dr. Subar commented that biomarkers of intake would be very helpful. An audience member commented that doubly labeled water appears to quantify underreporting. However, she asked about the validation of this technique and expressed concern about whether known dietary intake is actually underreported to the extent currently suggested by doubly labeled water studies. Dr. Subar indicated that the doubly labeled water methodology is well established as a measure of true energy expenditure in individuals, but she did not know if the intake matches the estimation in a steady state.

Another participant suggested that statistical modeling depends on the assumptions used. The assumption that a yearly intake reflects usual intake may be appropriate in some cases but not others, specifically in developing countries. Dr. Subar commented that the usual intake distribution is based on usual intake in the population. The participant suggested that in the United States, the intake does not vary much with the seasons, but in other countries seasons have considerable impact.

One question was raised about using the usual intake distribution when dealing with ULs. Dr. Subar was unfamiliar with any studies or deliberations intended to explore this particular issue. Another question was asked about the trustworthiness of the nutrient values on nutrition labels. Dr. Subar responded that others with expertise in this area would be better suited to answer the question.

HIGHLIGHTS OF OTHER IMPORTANT ISSUES: PHYSIOLOGICAL, ENVIRONMENTAL, AND GENOMIC FACTORS

Presenter: Cutberto Garza

Physiological, environmental, and genomic issues relate to the DRI conceptual framework as well as to the applications of the DRIs. This presentation first outlines some general principles to provide a context, then focuses on examples of challenges that physiological, environmental, and genomic issues present.

General Principles

The governing principle in any expanded consideration of physiological, environmental, and genomic issues is the definition of nutritional health. It is helpful to think about nutritional health in terms of a progressive overlapping continuum, moving from the bottom to the top of the trapezoid shown in Figure 3-12. The bottom of this continuum focuses on essential food components that, when lacking, give rise to unambiguous pathology related to a specific deficiency; or, if they are in excess, to an adverse effect. The single-agent, single-outcome paradigm governs this part of the continuum. Moving up along this continuum, there is a greater focus on primary and secondary prevention of nutrition-related chronic diseases. The top of the continuum is increasingly attentive to enhanced performance through improved nutrition.

Not surprisingly, uncertainty increases as we progress through this continuum from bottom to top. These uncertainties are due to decreases in basic knowledge (shown to the left of the trapezoid), reflecting the need for more research as we move from basic pathology and specific deficiency to concerns such as enhanced performance. There is also growing complexity of underlying biological mechanisms as we move toward enhanced performance. All this requires some broadening in the use of our tools. There is

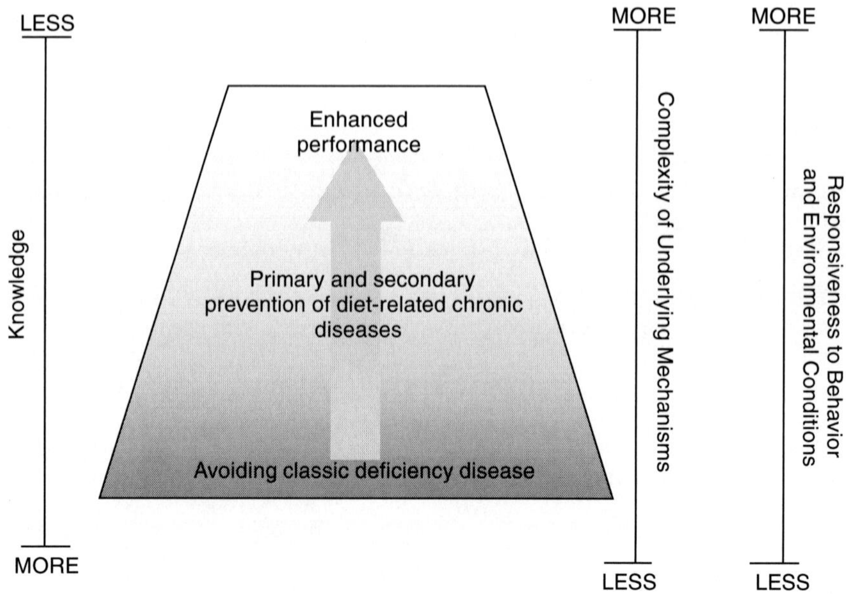

FIGURE 3-12 Nutritional health continuum.

also rising sensitivity to a wide range of behaviors and environmental conditions as we move from bottom to top. The significance of physiological, genomic, and environmental factors will differ along this continuum in ways likely to be specific to individual nutrients and life stages (Figure 3-12).

Two principles will help determine when such expanded considerations are appropriate. The first is that the anticipated benefit of modifying a reference value on the basis of any factor—physiological, environmental, or genomic—must be qualitatively significant to either individual or population health and well-being, somewhat analogous to, but the mirror image of, hazard characterization. Second, the equivalent of an individual- or population-attributable benefit must be quantitatively significant.

These seemingly straightforward statements beg the question of what triggers quantitative and qualitative significance. Criteria for determining qualitative significance are not independent from criteria for quantitative, and neither are likely to be determined purely on an objective basis. Assessments of both will be influenced by culturally or socially bound values and the ability to use the information.

Physiological Factors

Physiological factors include gender, age, reproductive status (including lactation), and body size. Considerations of body size are generally limited to expressions of nutrient needs per kilogram of body weight. Body size also incorporates elements of body composition to the extent that these two variables are related in a given population.

Four challenges exist with respect to physiological factors, recognizing that historically nutrient-based dietary recommendations have historically excluded nonhealthy populations:

1. The prevalence of obesity and overweight
2. The aging of the North American population
3. The increasing understanding of long-term risks associated with intrauterine growth retardation (slow-for-gestational-age infants)
4. High rates of prematurity, the health consequences of this condition, and increasing technological capabilities that enable survival at progressively lower gestational ages, which will bring special pressures to the DRI process

The IOM undoubtedly will be faced with including one or more of these conditions in the future DRI process; there may be a need to develop an ancillary effort to consider these groups beyond the brief paragraphs that have been included in the sections of the DRI reports labeled "special considerations." In addition, metabolic and other common morbidities that accompany overweight, obesity, aging, intrauterine growth retardation, and/or prematurity likely will influence recommended intakes, at least for some subgroups. The most salient example is the growing prevalence of Type II diabetes, which will be more difficult to ignore.

As challenging as these projections may seem, they pale when compared with the implications of considering environmental and genomic issues.

Environmental Factors

The framework that we have been using generally ignores environmental influences, with the possible exception of energy requirements. However, on an international level it is not uncommon, for example, to at least consider higher rates of endemic infectious diseases in the determination of nutrient requirements. Such conditions are often environmentally driven. Perhaps it is time that we too consider somewhat analogous environmental issues within our North American context.

Two examples illustrate this point. The first is the food environment. The millions of North Americans categorized as overweight or obese did not plan to develop these conditions. For the most part, overweight or obesity happens. Although one has to think intentionally about being healthy, consumers do not have to be as intentional about becoming overweight. Are there inherent biological reasons why health could not also happen to people as unintentionally as overweight or obesity appears to occur?

Consumers experience free market forces related to food to a much greater degree than we appear to tolerate other areas of public health and safety. For example, given the perils of unsafe highways and cars, we do not rely solely on educating the public so that they can become better drivers: We engineer safer highways and cars. Are there analogous roles the DRIs can or should play to help safeguard nutritional health, such as modifying the width of the Acceptable Micronutrient Distribution Range, or do the DRIs make sense at all without greater specificity in terms of the type of fat?

A second dimension of the food environment is our increasing ability to manipulate nutrient intake through fortification, genetic engineering, and supplements. The potential for adverse nutrient interaction merits continued close attention. Perhaps the most salient example of the importance of such considerations is higher than initially projected levels of folate intake and their potential adverse impact on individuals and groups with inadequate vitamin B_{12} intakes and/or impaired vitamin B_{12} uptake capabilities or the progression of early cancer.

The second example relates to environments that either enable or discourage physical activity. Although we think of physical activity primarily in terms of weight status, physical activity also influences the risk of other chronic conditions. Heightened consideration probably will focus on whether nutrient needs are modified by diverse levels of physical activity if chronic disease risk reduction is among the desired outcomes.

Genomics

In terms of issues that fall under the broad category of genomics, current considerations for DRI development are limited to genomic variability, which specifically takes the form of including body size (to the degree that size is genetically controlled) in estimating requirements. The broad considerations of interindividual variability may be considered, and there may be attempts to address a few specific polymorphisms.

Until recently, other than for folate, no other adjustments were made for well-known polymorphisms. Generally, nutrient needs modified by groups of specific polymorphisms were viewed as condition requirements. Among the most salient examples of these are vitamin D-dependent rickets, hemochromatosis, and phenylketonuria, conditions that either increase or decrease appropriate levels of nutrient intake.

What about the future? For the most part, complex traits that account for diet-related chronic diseases appear to be influenced by multiple polymorphisms that individually have only modest adverse or beneficial effects on risk, but collectively appear to have significant influence. A study reported recently in the *New England Journal of Medicine* (Rosenzweig, 2007) that used genomic scanning techniques to assess coronary disease risk supports this view. The value of such work in improving the definitions of risk, enhancing mechanistic understanding, and generating potential interventions for future investigation is acknowledged. For the moment, however, results of such studies appear to have limited immediate impact on specific preventive measures.

We also have to recognize that work such as that from Waterland and others (e.g., Waterland and Jirtle, 2003; Waterland et al., 2006) points to the complex epigenetic effects of some nutrients in determining phenotype. There is little doubt that greater understanding of environmental–genomic interactions and the influence of genomic context will result in improved definitions of risk and mechanistic underpinnings.

Also, based on what we know now, it is likely that improved understanding of these relationships eventually will result in better individualized care. What is less clear is how this type of information will help in designing strategies that target populations, particularly as North American societies become more ethnically diverse.

Implications

Future approaches for DRI development are likely to be increasingly more sophisticated in their inclusion of an array of physiological, environmental, and genomic characteristics. The interplay of these factors in determining the prevalence of various phenotypes will need to be recognized, and the interpretation of the special nutrient needs imposed by this interplay will require an expanded DRI process. This increased sophistication will impose important challenges to further address knowledge gaps, mechanistic complexity, and the present inadequate understanding of interactions among diverse environmental conditions and individual behavioral choices.

Finally, an improved understanding of genomic influences on health will cause us to rethink the use of DRIs in designing strategies to promote individual and population health.

General Discussion

An audience member commented that genomic variability and the presence of polymorphisms will undoubtedly play an increased role in DRI development. However, after describing the example of methylenetetrahydrofolate reductase polymorphism, he suggested that the changes involved may not be dramatic. Dr. Garza added that we often forget that these polymorphisms were positively selected. At some point in our evolution, they must have played some beneficial role. In some context, they may increase risk, whereas in other contexts, they may be protective.

Another participant addressed the issue of environmental influences, noting that Dr. Garza had mentioned infectious diseases as pertinent to nutrient reference values for persons in developing countries. Given that inflammation is shown to play a role in the pathogenesis of chronic diseases and may be relevant to the aging North American population, the participant questioned whether inflammation should be added to the list as either a physiological or environmental factor to be considered. Dr. Garza responded that aging is germane, and the physiological adjustments and metabolic abnormalities that accompany aging, are relevant to the derivation of future DRIs.

4

General Guidance for Users of DRIs: Session 3[1]

Session 3 participants were asked to develop their presentations keeping in mind several questions: Is the general guidance for users of Dietary Reference Intakes (DRIs) appropriate as well as consistent with the purpose and goals of the DRIs? What more needs to be done? In addition, specific topics of interest included: How should Adequate Intakes (AIs) be used for planning and assessing, especially within the context of the total diet? Can clarification be provided on the differences between groups and individuals and between applications for small groups and those for large groups? What "practical guidance" and tools can be provided to assist practitioners?

The moderator of this session was Mary Bush of Health Canada. The session opened with a presentation by Dr. Christine Taylor, the Study Director, on the wide-ranging issues surrounding general guidance for users. This was followed by a joint presentation by Dr. Suzanne Murphy of the University of Hawaii and Dr. Susan Barr of the University of British Columbia on the issues and options for enhanced guidance in terms of planning and assessing the total diet. The final presentation was given by Dr. Valerie Tarasuk of the University of Toronto, who discussed issues related to a framework for individual- and group-level applications.

Discussions open to all audience members were held after each presen-

[1]This chapter is an edited version of remarks presented by Drs. Taylor, Murphy, Barr, and Tarasuk at the workshop. Discussions are composites of input from various discussants, presenters, moderators, panelists, and audience members.

tation. Panel discussions exploring how guidance for users of DRIs could be enhanced closed the session.

OVERVIEW: ISSUES RAISED ABOUT GENERAL GUIDANCE FOR USERS

Presenter: Christine Taylor

The issues surrounding general guidance for users are, overall, both wide ranging and overlapping. The purpose of this presentation is to offer some context for this session's discussions.

It is important at the outset to highlight at least two different interpretations of the term "uses" as they relate to DRIs because these interpretations have caused confusion. Some refer to the "uses of the DRIs" to mean their general purpose and intent, consistent with the conceptual underpinnings of DRIs. Others refer to "uses of the DRIs" when they are referring to the specific guidance for assessing and planning for individuals and groups. While these are not necessarily disparate, they each have a different focus. The purpose of DRIs is an important topic that properly belongs with those discussions relevant to conceptual underpinnings that occurred in Session 1. The discussions for this session on general guidance for users relate to the approaches outlined for applying the various DRI values to accomplish certain tasks.

From the 1940s through the late 1980s, the Recommended Dietary Allowances (RDAs) and Recommended Nutrient Intakes (RNIs) were issued with little or no guidance for users. With the 1994 expansion to a more complex set of reference values, the need to provide general guidance was evident. This guidance was offered in two publications: *Applications in Dietary Assessment* and *Applications in Dietary Planning* (IOM, 2000a, 2003a). These documents were the work of the Subcommittee on Interpretation and Uses of Dietary Reference Intakes.

The starting point for the guidance was the general categories of tasks commonly carried out using a nutrient reference value: assessing and planning dietary intakes for groups and individuals. These activities are often illustrated using a two-by-two table, shown as Figure 4-1.

In essence, as outlined in the two guidance publications, the DRI value to be used will differ depending on which activity is being carried out. For instance, in the case of guidance for assessing groups, the focus is on using the Estimated Average Requirement (EAR) (not the RDA), cutpoint methods, and probability approaches. For guidance for planning for groups, the goal is identified as a low prevalence of inadequate intakes and consideration is given to definitions of acceptable prevalence of inadequate intakes. For assessing individuals, the guidance contains both qualitative and quan-

Assessing Groups	Planning for Groups
Assessing Individuals	Planning for Individuals

FIGURE 4-1 "Two-by-two table": General applications of Dietary Reference Intakes.

titative discussions. Guidance for planning for individuals emphasizes that the use of the RDA is preferable to the use of the EAR; also, AIs can be used in place of the RDA, but there is a greater level of uncertainty. This guidance derives from a statistical foundation and is based on understandings of distributions, normality, and probability.

Whether the concept reflected in the two-by-two table works both conceptually and during the implementation process has been questioned. Some have asked whether there are special issues if a program fits within more than one box of the two-by-two table. Others have noted that guidance for planning seems to have been less well implemented than guidance for assessment.

Broader questions for this session's discussions include the following: Given the expansion of the DRI values and the subsequent development of guidance for their use, what have we learned about the needed guidance? What are its pros and cons? What, if anything, needs rethinking, further elaboration, or more work? How would changes in the conceptual framework for DRIs that have been suggested at this workshop impact DRI guidance in the future?

A helpful set of background papers was developed for this workshop, targeted specifically to guidance for users. The U.S. government, Health Canada, the American Dietetics Association, and Dietitians of Canada all offered input on the topic. Numerous issues were highlighted in these documents, but they generally fell into two categories:

1. Further methodological work, gaps, and conceptual evolutionary changes that need to be addressed or need some type of revamping
2. A set of questions about ease of use, practicality, and the need for simpler guidance

Specific questions raised included what further work needs to be pursued to address the emerging world of statistical methods and their application to DRI guidance for users (particularly individuals), and to what extent guidance is limited by lack of research on relevant methodologies.

In gathering input on issues related to guidance for users, two special challenges emerged. One is that the DRIs have to operate in the context of the total diet. They are created for individual nutrients, but when the user applies them, it is done in the context of the total diet. One snag is that for some nutrients they are using EARs, whereas for others they are using AIs. A second special challenge surfaced relative to the framework for individual- and group-level applications. It was asked whether it was worthwhile to distinguish between these and how their distinctions could be made clear.

Regarding practitioners such as dietitians who work in a wide variety of settings in which DRIs are used, the main theme voiced was the difficulty in understanding DRIs (separate from applying the DRIs). Practitioners are interested in more clarity, more "practical" guidance, more guidance on use with individuals, and more tools, such as software. An interest in "simplicity" was expressed.

As mentioned earlier, the issues for this topic area are quite diverse, and they provide a rich background for the session's discussions. The overall questions are the following: Is the guidance heading in the right direction? Is it consistent with the purpose and goals of the DRIs? What more needs to be done?

DISCUSSION

Discussant: Johanna Dwyer

The session moderator, Ms. Bush, introduced Dr. Dwyer and invited her to offer an opening remark.

Discussant Opening Remarks

Dr. Dwyer mentioned several practical issues concerning the use of the DRIs. The first focuses on the level of precision needed for assessment and planning for individuals. The precision needed may be more than for food-based guidance, but less than the prescriptive recommendations needed in medical nutrition therapy. A second concern is harmonizing the DRIs, food labels, and MyPyramid advice with respect to chronic degenerative disease risk, especially when the focus is on counseling individuals. A third issue is the application of DRIs to those with treated diseases (e.g., those taking high blood pressure medication) and who are apparently healthy. For most nutrients, the DRIs may be appropriate for these persons, with the exception of specific nutrients that may be affected by the disease in question.

Dr. Dwyer also suggested that assessment and planning require estimates of *total* nutrient intakes from all sources, including fortificants and dietary supplements. The ability to link intake with health consequences depends on usual total dietary intakes from all sources of nutrients, not just food. However, estimating the distribution of total intake for a population is challenging, and current estimates of intake suggesting low or high intakes in North America may be questionable. Are these findings due to uncertainties in estimating the requirements, difficulties in obtaining accurate dietary intake information, the unintended consequences of applying the values incorrectly, or some other factor?

Finally, some of the problems experienced by users may be addressed if the DRI process incorporated systematic evidence-based reviews (SEBRs) on specific questions and a risk assessment approach. However, the lack of data may be challenging. Regarding the development of future DRIs, it will be vital to identify the most critical gaps in knowledge and methodologies and to obtain a broad consensus on how best to resolve these. However, this should not justify endless procrastination.

General Discussion

Dr. Taylor joined Dr. Dwyer on the dais. When asked for any initial thoughts, Dr. Taylor responded that there seemed to be an emerging theme that practitioners need assistance in understanding and applying DRIs, and that a starting point may be educational efforts. Dr. Dwyer added that *Dietary Reference Intakes: The Essential Guide to Nutrient Requirements* (IOM, 2006) was accessible to all. She also noted an online course produced by Dietitians of Canada as well as an array of helpful articles. She suggested that the Institute of Medicine (IOM), Canadian and U.S. governments, and professional societies should partner with various user groups, such as dietitians and family practitioners, to develop additional tools. In short, there is a need to train practitioners to interpret values rather than blindly using reference values.

Meaning of "Uses"

An audience member remarked that the term "uses" seemed to have various meanings. "Uses" in the context of the DRIs is a generic term that focuses on planning and assessment for the general population. However, the workshop participants seem to be referring to other "uses"; for example, are these DRIs to be used for fortification or for food programs? She queried whether, in the current context, Dr. Taylor was referring to uses generically or seeking information that is much more specific in terms of its application. Dr. Taylor responded that the focus of this session was general guidance for users in the context of planning and assessment. She added that the overall perspective is that specific applications that require more in-depth or focused scientific considerations would be best addressed on a case-by-case basis, perhaps by special study committees if the IOM were to be involved. She commented, however, that the developers of this workshop had also noted the various meanings associated with the term "uses" in the context of DRIs. Another audience member commented that during her work on the United Nations University Harmonization Report, there had been a vigorous discussion of the uses of dietary standards and recommendations. It became clear that there were two ways of looking at "uses": one way is the applications of the DRIs to specific programs, and the second is the methods for planning and assessing total diets for groups and individuals.

A comment was made that application of the risk analysis paradigm to DRI development elevates the question of "uses." She suggested it will frame the set of issues important to sponsors and other stakeholders.

Chronic Disease Considerations

A participant asked whether planning for groups should return to focusing solely on non-chronic disease endpoints or whether it should continue to embrace the chronic disease question. Examples of the challenges included the inclusion of macronutrient recommendations and tolerable upper intake level (UL) considerations in planning school lunches and establishing the basis for nutrition labeling. Dr. Dwyer responded that in planning for groups, it may be better to split adequacy and chronic disease into two reports. She added, however, that eventually they must come together in a single recommendation because a population is subject to both risks.

Simplification

A participant issued a word of caution about simplification, suggesting the issues are not simple. While the most common user may be private practitioners, their use of the DRIs may not result in proper application of values. It is more important to ensure that DRIs are used correctly. He suggested it may be better to rank the importance of the usage rather than the frequency.

SPECIAL CHALLENGES: PLANNING AND ASSESSING THE TOTAL DIET—WHAT ARE THE ISSUES AND WHAT ARE THE OPTIONS FOR ENHANCED GUIDANCE?

This presentation is divided along the lines of the paradigm, into assessing and planning intakes. Several crosscutting issues apply to both.

Assessing the Total Diet

Presenter: Suzanne Murphy

To use the DRIs to assess usual intake for an individual, one needs to know the probabilities of inadequacy and excess. For a group, one needs to know the prevalence of inadequacy within the group and the prevalence of intakes at risk of being excessive. Although simple in concept, this application is difficult in practice because of three challenges.

Challenges and Possible Solutions

The first challenge is associated with combining AIs and EARs. For most nutrients, the EAR and RDA are only about 20 percent apart. Generally, as shown in Figure 4-2, in which the left-hand curve is drawn more steeply than usual to reflect the assumption, the AI is assumed to lie beyond the RDA, but is potentially slightly below the RDA. It is a vaguely defined number and, in many ways, not very useful.

For nutrients with an EAR, the probability of inadequacy for an individual as well as the prevalence of inadequacy for groups can be calculated. For nutrients with an AI, neither probability nor prevalence can be

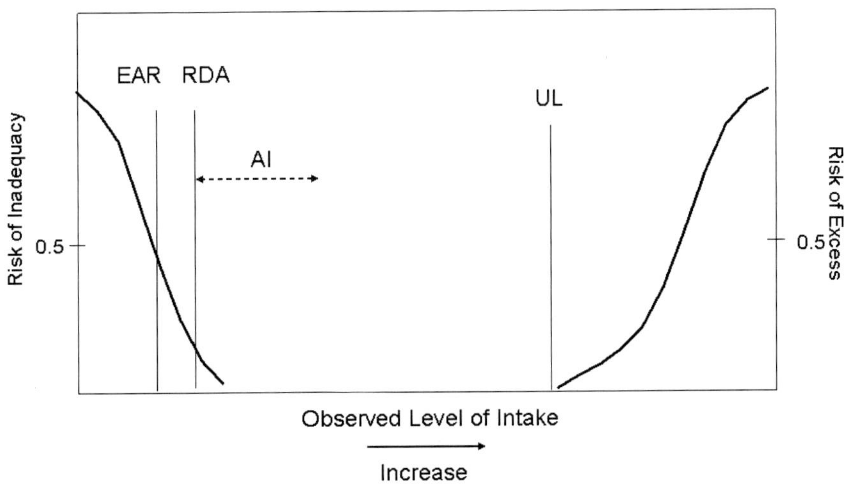

FIGURE 4-2 Relationship of the AI to the EAR and RDA.
NOTE: EAR = Estimated Average Requirement; RDA = Recommended Dietary Allowance; AI = Adequate Intake; UL = tolerable upper intake level.

estimated; it can be stated only that intake at or above the AI should have a low probability or a low prevalence of inadequacy.

The reference value for calcium is an AI. In practice, one option for estimating the probability of inadequacy is to assign approximate probabilities for a range of intake levels, based on an evaluation of the data used to set the AI. For example, the following probabilities shown in Table 4-1 might be used to assess persons over 50 years of age for calcium.

Someone with an intake of calcium of less than 300 mg/day probably has nearly 100 percent probability of inadequacy. Someone whose intake is at the AI of 1,200 mg/day should have a very low probability of inadequacy.

A second challenge relates to the assessment of individuals. Theoretically, the probability of adequacy or inadequacy for an individual can be calculated using the distribution curve and considering the standard deviation (SD) of the requirement. However, the methodology recommended in the IOM assessment report (IOM, 2000a) was to calculate a confidence of adequacy. That is, instead of comparing the intake of the individual with the EAR and its SD to get the probability, the SD should be increased to consider the number of days of intake and the within-person variance of intake. The resulting confidence of adequacy is typically less than the probability of adequacy.

Using thiamin as an example, suppose that the average intake for an individual over 3 days is 1.3 mg. The thiamin RDA for an adult is 1.2 mg, which is below this person's actual intake. If this was the person's usual intake, the probability of adequacy would be about 98 percent. However, if the probability of adequacy is adjusted for the number of days and the within-person variability in thiamin intake, the confidence of adequacy is only 85 percent. This value is difficult to interpret because it does not necessarily mean a person should increase his or her intake. Rather, it probably means the number of days of intake data should be increased. By combining these two concepts, the confidence of adequacy often cannot be interpreted. A better method would be to calculate both the probability of adequacy and the confidence of adequacy.

TABLE 4-1 Examples of Approximate Probabilities That May Be Used to Estimate Probability of Inadequacy

Calcium Intake (mg/day)	Probability of Inadequacy (percent)
0–300	Probability = 100
301–600	Probability = 75
601–900	Probability = 50
901–1,200	Probability = 25
Above 1,200	Probability = 0

Alternatively, a qualitative evaluation of the individual's intake can be performed. If data on several days of intake show the usual intake is below the EAR, an increase in intake is probably needed. If the intake is between the EAR and the RDA, it probably needs to be improved. If it is at or above the RDA, it is probably adequate.

A third challenge focuses on the need for better communication of the concepts associated with "uses" of the DRIs. For example, separating dietary collection methods from dietary evaluation methods would be helpful. The term "uses" should be called "evaluating intakes," and would include guidance on calculating the probability of inadequacy for individuals and the prevalence of inadequacy for groups using the EAR as well as information for correctly using other DRI values (e.g., the Estimated Energy Requirement, AI, UL, Acceptable Macronutrient Distribution Range [AMDR]). Collecting intakes would cover the choice of methods used to collect the data, the number of days of data needed, and adjusting the data to reflect usual intake distributions.

Moreover, the concept of uses should be separated from applications. Uses can be considered as the theoretical applications of the DRIs, or how the DRIs are used to evaluate and plan intakes. The term "applications" should be saved for the ways in which people apply the DRIs for specific problems (e.g., dietary counseling, food labels, Special Supplemental Nutrition Program for Women, Infants and Children [WIC] vouchers, dietary guidelines). Each type of problem might be better addressed by its own study committee and report. Dietitians also need a separate report on how to use the DRIs for dietary counseling; the American Dietetics Association and the Dietitians of Canada should consider collaborating on such a report.

Implications

General guidance on uses of the DRIs should focus only on evaluating and planning intakes. A separate group should consider dietary collection methods, and separate reports might be written for various applications of the DRIs (as was done for the WIC package revisions and the changes to the food labels). Collaborative groups, as well as IOM study committees, could address some of these topics.

Planning the Total Diet

Presenter: Susan Barr

Given that the goals of dietary planning for individuals using DRIs are to achieve intakes with a low risk of prevalence of inadequacy or excess, the suggested approach is fairly straightforward. Unless there are special considerations (e.g., vitamin C for a smoker or iron for a vegetarian marathon runner), the planning goals can be met by intakes that meet the RDA or AI, are below the UL, meet energy needs, and are within the AMDR. These goals have been operationalized using food guides.

In contrast, proposed approaches to planning for groups are considerably more complex, depending on whether the groups are homogeneous or heterogeneous in terms of energy and nutrient requirements, whether there are vulnerable subgroups that can be identified and targeted for intervention, and whether the requirement distribution is skewed or symmetrical.

The simplest case is planning for a homogeneous group with a symmetrical requirement distribution (e.g., vitamin C for nonsmoking young men). Planning in this case is accomplished by the EAR cutpoint method. The goal of a low prevalence of inadequacy is met if the proportion with usual intake below the EAR is minimal. If the baseline assessment has revealed a high prevalence of inadequacy, as estimated by the proportion below the EAR, planning would be done to shift the distribution upward to minimize the prevalence of inadequacy. Because variability in intakes is generally greater than variability in requirements, the mean or median intake of that distribution with the low prevalence of inadequacy would invariably be above the RDA. At this point, however, it is not known whether implementing this approach by changing the foods offered would actually shift the distribution or simply change its shape, so the approach remains largely theoretical.

Key Considerations and Challenges

The methods described in the IOM planning report (IOM, 2003a) focused on planning for single nutrients, whereas planners typically need to plan total diets. This process presents a number of challenges:

- One of the first challenges is "where to start" when planning for a total diet versus a single nutrient. This will vary depending on whether there is a preexisting set of menus and what the results of a baseline assessment show. If there are preexisting menus and the assessment showed a high prevalence of inadequacy for only one or two nutrients, it would be logical to start the process by planning to modify intakes of those specific nutrients. If, however, planning were to occur de novo (e.g., for a new retirement facility for 300 seniors of Chinese ancestry), more guidance would be helpful. Establishing the energy "budget" within which nutrients need to be incorporated would be a logical starting point, but other approaches could be considered.
- A second challenge relates to the need to establish priorities in situations where not all the planning goals can be met. This situation is

most likely to arise when planning for those with low energy needs, such as seniors in a retirement facility. However, it also arose in developing Canada's Food Guide, where it was not feasible to reach targets for linoleic acid, potassium, vitamin D, and fiber, among others. Considerations in setting priorities might include whether to focus on nutrients that have an EAR/RDA rather than an AI, whether to focus on nutrients that have more direct consequences of inadequacy, and whether supplementation or fortification is feasible.

- Third, planning must sometimes occur for nutrients for which no specific guidance has been provided. This is the case for saturated fatty acids, *trans* fatty acids, and cholesterol, which lack ULs. Although the rationale for not having a UL is clearly explained in the macronutrient report, users are left with the need to pick their own benchmarks or use guidance from other groups.
- Fourth, "conflicting" DRIs must be addressed. Several macronutrients have an EAR/RDA (or AI) and an AMDR. For fatty acids, the AIs for the N3 and N6 fatty acids were based on median intakes of the U.S. population, while the AMDR was set with the AI as the lower boundary. For nutrients with an AI, one normally plans for a low prevalence of inadequacy by planning for median intakes to equal the AI. However, if this were to be done for N3 or N6 fatty acids, 50 percent of the group would be below the AMDR. It is difficult for users to know how to proceed in such a case.
- Fifth, it should be recognized that food guides traditionally have been used for individuals (by being designed to meet the RDAs), but could potentially be used for groups as well if they were developed with that objective. Canada's Food Guide was designed to lead to a low prevalence of diets with nutrient levels below the EAR, to meet most AIs, etc. This means it could be used as a starting point to plan diets for groups as well as for individuals (although, as indicated above, it was not possible to meet all planning goals).

Although the approach used to develop Canada's Food Guide was based in part on the planning paradigm outlined in the planning report, it did not use a baseline usual intake distribution as the starting point. Instead, it was based on the assumption that intakes would comply with the food guide. If menus were planned using the food guide and a high prevalence of inadequacy was found after intakes were assessed, planners would still need to try to shift the distribution upward.

Implications

Users have faced many challenges in adopting and understanding the new paradigm for planning for total diets. One is knowledge gaps, which need to be addressed. Although resources have been developed to help users, the material is not easy to understand and requires a considerable investment of time, energy, and, sometimes, money. The result has been some inappropriate uses of DRIs in dietary planning.

Another challenge is issues related to data and software. For example, before using this approach, a dietitian working in an assisted living facility would need to obtain 24-hour recalls from the residents, obtain repeat recalls on a subsample, analyze the food records to derive nutrient intakes, and then use sophisticated software to obtain the usual nutrient intake distributions. This would require access to software and the time and ability to the software.

There are also many gaps between theory and practice. Because DRIs are intended for use with healthy individuals, one gap is how to plan for those who are not healthy. Of greater importance, however, is the question of whether the concept on which the group planning paradigm is based (i.e., "shifting distributions") works in practice; to date, there are few examples of successful use of this approach. Clearly the DRI framework has allowed good progress to be made, but many issues still need to be resolved in terms of guidance for users.

DISCUSSION

Co-Discussants: Patricia Guenther and Krista Esslinger

The session moderator, Ms. Bush, introduced the discussants and asked them to offer opening remarks.

Discussant Opening Remarks

Dr. Guenther noted that the U.S. Department of Agriculture's (USDA's) Center for Nutrition Policy and Promotion uses the DRIs when developing guidance for consumers. The food patterns that make up the MyPyramid Food Guidance system aim to meet the RDAs, ULs, and AMDRs, plus additional standards from the *Dietary Guidelines for Americans*. She suggested that practitioners who develop menus or plan diets should rely on government food-based recommendations rather than using the DRIs directly. Those recommendations have already incorporated DRI translations into useful food-based recommendations. She noted that such guides are relevant in most situations and can be modified for people with special dietary needs—for example, by modifying the published food group composites underlying MyPyramid.

Regarding the USDA experience with DRIs, Dr. Guenther noted that her agency's efforts to use RDAs, AIs, ULs, and AMDRs for planning diets of individuals had been successful, as had the efforts of the USDA Agricultural Research Service in using EARs to assess the diets of populations. However, there were challenges in carrying out the other applications—assessing diets of individuals and planning diets for groups—because the statistical methods needed were not fully developed within the IOM guidance documents. The confidence of adequacy approach for assessing individual diets seemed impractical, and the probability of adequacy approach would require knowledge of an individual's long-term nutrient intake, which is impossible to measure accurately. Other USDA efforts demonstrated that the guidance relevant to group planning was impractical because the shapes of the target nutrient intake distributions were difficult, if not impossible, to determine in an environment where intervention is intended to change the distribution of the nutrient intake. She noted that the methodology used to develop Canada's Food Guide is a more appropriate way to use the DRIs to plan and assess group and individual diets. To be fully effective, however, it requires EARs for all nutrients of public health concern. The EARs and ULs are the essential DRIs, and the AI should be reincarnated as an EAR that is set with a demarcation of greater uncertainty.

Ms. Esslinger remarked that although nutrient reference values are an important scientific input to dietary assessment and planning, they are not the only input. A "diet" comes about by considering nutrients as well as food selection, including environmental and cultural contexts in which the food is eaten. Therefore, when considering the total diet, it is most meaningful to speak in terms of foods. The translation of nutrient recommendations into food-based recommendations is where the challenge lies. The need to focus on foods rather than nutrients may be an explanation for why the IOM guidance for users has been used infrequently for planning. Ms. Esslinger also observed that the DRIs are not used frequently in clinical and dietetics settings for a variety of reasons. One important reason is that the current methodology recommended for individuals does not seem to be practical for assessment of the total diet. She suggested there is a need for tools that use the DRI values as well as other information about diet, and development of these should be the responsibility of users rather than IOM study committees.

Finally, in turning to experience in Canada, Ms. Esslinger noted that the EAR cutpoint method was used with success when developing Canada's Food Guide. She said the nutrient contents that were the most difficult to achieve when creating the food intake pattern were for those nutrients with a large discrepancy between the DRIs and actual intakes. For example, nutrients with an AI based on the highest quintile of intakes (e.g., fiber and potassium) were much more challenging than those with an AI based on median intakes (e.g., essential fatty acids).

General Discussion

DRIs in Clinical Practice

In the opening discussion among the presenters and discussants, one presenter emphasized the desirability of convening a working group of dietitians and others to discuss DRI application in clinical practice. She pointed out that the background papers by the Dietitians of Canada and the American Dietetic Association indicated that dietitians want to use the DRIs, but are uncertain how to apply them. She suggested that dietitians do not just use food guides; some situations require nutrient reference values, such as when working with a population that has specific nutrient needs, so they must be able to adjust the nutrient intake recommendations. A second presenter agreed, but a discussant disagreed, saying that when she was a clinical dietitian counseling patients with a need for dietary modification (e.g., someone with renal disease), she did not tell her patients how much of the various *nutrients* to consume, but instead suggested the amounts and types of *foods* they should consume.

Interface Between Individuals and Groups

When the discussion was opened to the audience, a participant commented that the developers of the DRI guidance for users failed to recognize the interface between individuals and groups. Existing guidance indicates that one plans for the individual using the RDA and assesses plans for a group using the EAR. This means, on one hand, that each individual has to be seen at low risk and, on the other hand, that the population at low risk is a goal. These are very different concepts. That clash between individuals and groups needs to be resolved. In response, one of the speakers indicated that if an individual is at the RDA, by definition he or she has a 2–3 percent risk of not meeting the individual requirement. For example, in a population of 1,000 individuals, all of whom had an intake equal to the RDA, there would be a 2–3 percent prevalence of inadequacy. No one in the population would be below the EAR, so the cut point method would not be applicable in that situation because there would be no variability in intake. The original questioner disagreed with these calculations,[2] but it was noted that the issue would be further covered in the talk by Dr. Tarasuk on individual- and group-level applications.

[2]The questioner later explained his disagreement as follows: Given the impossible situation of a group of 1,000 individuals each consuming an intake intended to cover all but 2.5 percent (risk 0.025), the expected prevalence of inadequacy would approximate 0, and thus the risk to a randomly selected individual would be approximately 0, not 0.025.

Food Supply Changes and Intake Distribution

An audience member noted that one of the speakers said there were no data concerning the distribution of intake when the food supply changes. She suggested it would have been possible to use intake data published before and after iron and folic acid fortification—showing both a shift and a skew in distribution—as well as data from studies that evaluated intakes before and after the introduction of food stamps. In response, the speaker indicated that most of the studies referred to by the commenter were published after the work of the presenters. In addition, the published papers generally did not show percentiles, which are needed to determine how the shape of the distribution changed.

Skewed Distributions

An audience member remarked on the conceptual advances and new tools offered by the Subcommittee on Interpretation and Uses of Dietary Reference Intakes. However, he highlighted a problem in the approach for predicting the shape of the intake distribution given an intervention that shifts the distribution—the assumption that there will be a shift *and* no skew when the intent of the intervention is to skew the distribution results in a target value well above the RDA for many nutrients. He suggested this will cause problems in planning diets for group feeding situations, and work to address this is needed.

Other Comments

A participant emphasized the challenges in developing a food guide based on the DRIs while maintaining recommended caloric intakes. Another commented that the U.S. MyPyramid and Canada's Food Guide are both useful, but there is also a place for the use of food exchange lists and other tools. Another asked who is responsible for developing software that would allow practitioners to make better use of DRIs. A discussant responded that more consensus on the methodology is needed before more software is developed. A final commenter focused on an earlier suggestion to develop an AI-based approach to examine the probability of inadequacy. She noted that a careful read of the DRI reports could reveal the point at which one sees an indicator of adequacy in a group (albeit not necessarily an indicator used to create an EAR) and through that process find a way to establish cutoff values.

SPECIAL CHALLENGES: WHAT ARE THE ISSUES RELATED TO A FRAMEWORK FOR INDIVIDUAL-LEVEL AND GROUP-LEVEL APPLICATIONS?

Presenter: Valerie Tarasuk

The two-by-two table that divides DRI applications into assessment and planning for individuals and groups (see Figure 4-1) is the organizing framework for the discussions that appear in the DRI applications documents (IOM, 2000a, 2003a). Four core concepts underlie the framework and are articulated in these documents:

1. Requirements are recognized as distributions, whether or not they can be mapped.
2. If the distribution of requirements is known, there is a known probability of inadequacy or adequacy associated with each intake level in relation to that distribution.
3. The relevant point of comparison for nutrient requirements is usual dietary intakes.
4. Planning—either for the usual intake with a low risk of inadequacy for an individual or for a distribution of usual intakes with an acceptably low prevalence of inadequacy for groups—is fundamentally an extension of assessment.

Challenges Associated with Implementing the Core Concepts

Implementation of these concepts has presented challenges. One challenge is that the framework functions only where probability theory can be applied. Therefore, we cannot apply AIs, AMDRs, or ULs using the framework (as we can for EARs and RDAs). These values do not allow us to differentiate between individuals and groups or between assessment and planning. Their application is arbitrary. Moreover, there is no application guidance for report recommendations such as "saturated and *trans* fat intakes should be as low as possible."

A second challenge is that the strength of the assessment paradigm is at the population level. Assessment of the adequacy of an individual's usual intake is severely limited by our inability to assess that intake with any degree of precision. A related issue is the questionable applicability of the framework to small groups. Assessment and planning for small groups hinge on our ability to estimate distributions of usual intake; factors to be considered include sample size, number of replicate observations, and the representativeness of the subsample from which they are drawn. The reliability of the estimated prevalence of inadequacy is a function of the reliability of the estimated distribution of usual intakes. Part of the challenge is the difficulty in deriving stable estimates of usual intake when there are only a few observations and a small sample size. For example, our research found a prevalence estimate of 26 percent for folate in a small group of women (n ≈ 300, with replicate observations for 80), but the confidence intervals around that prevalence estimate ranged from 0 to 85 percent.

The other issue with small-group applications is the nature of those groups. They may include nursing home residents, the elderly, or other subgroups of the population that fall outside the core assumptions of DRI development. The appropriateness of applying the DRIs to those groups can be questioned. Assessment and planning activities for small groups would benefit from clearer guidance on the conditions under which group-level assessment and planning methods can be applied with reasonable confidence; alternative approaches to assess and plan for nutrient adequacy among small groups; and different reference standards for use in clinical settings.

The third challenge relates to the interface between populations and individuals. The two-by-two table (Figure 4-1) draws a firm line between individuals and populations. However, attempts to apply the DRIs, particularly for regulatory affairs and public policy purposes, show that this distinction is a gross oversimplification. The goal of planning for individuals is "to ensure that the diet as eaten has an acceptably low probability of nutrient inadequacy while simultaneously minimizing the risk of nutrient excess" (IOM, 2003a). We assume the RDA is a point on a distribution of requirements that lies at the upper tail, which will encompass the needs of 97 or 98 percent of the population. If someone is consuming a dietary intake at the level of the RDA, he or she will have a low probability of inadequacy.

Using the RDA as the goal of planning for an individual assumes that one will achieve a usual intake approximately equal to the RDA (i.e., there is no between-person variation in intakes), but this is not realistic for population-level applications. The midpoint of the distribution of usual intakes that achieves that low prevalence of inadequacy will be higher than the RDA. Thus, the goal in planning for groups is to achieve a distribution of usual intakes that has a low prevalence of nutrient inadequacy and a low proportion of intakes above the UL (IOM, 2003a). This takes into account between-person variation in usual intakes. The implication of these differences is that, as currently defined (IOM, 2003a), population- and individual-level approaches to planning are not interchangeable, but yield different nutrient targets with potentially different outcomes.

An Illustration of the Challenge

Examination of the U.S. and Canadian processes to update their food guidance by incorporating the DRIs provides an illustration of the current controversy surrounding the interface of individual- and population-level applications. The U.S. group, when faced with the task of reframing MyPyramid, decided that its purpose was to plan for individuals using the RDA. To develop food intake patterns, it calculated the nutrient profile for each food group or subgroup from the weighted average of the nutrient content of foods in the group, based on National Health and Nutrition Examination Survey (NHANES) food consumption data, then used an interactive process of modeling to determine the amounts from the food group composites required to meet nutrient and energy goals.

The purpose of Canada's Food Guide is "to assist the people of Canada in making food choices that promote health and reduce the risk of nutrition-related chronic diseases" (Katamay et al., 2007). The Canadian goal in the updating process was a low prevalence of diets with nutrient content below the EARs and a median nutrient content approximately equal to the AIs. They first modeled food composites to obtain a food intake pattern for each age and gender group with satisfactory average nutrient levels, similar to the U.S. group. However, once they obtained some food composites with reasonable nutrient values, they assessed the nutrient distributions from 500 simulated diets designed to comply with the test pattern for each age and gender group (i.e., recognizing variation in food selection). They then used an iterative process to identify the dietary pattern with a low prevalence of inadequacy and a median approximately equal to the AI.

Although these were two totally different approaches to the implementation of the DRIs for food guidance purposes, the results appear to have been similar (Table 4-2). Perhaps this can be explained by the facts that nutrients with AIs were treated similarly in both modeling exercises and that translation of multiple nutrient targets into numbers of servings from different food groups blurred the distinctions between the methods. Following either guide results in intakes well in excess of the RDAs for most nutrients. Although these results might suggest that the frameworks for individuals and populations are interchangeable, they will result in different nutrient targets in many applications.

TABLE 4-2 Food Guide Recommendations for a Sedentary Adult Woman ≤50 Years of Age in Canada and the United States

	Number of Servings	
Food Group	United States	Canada
Grain products	6	6
Fruit and vegetables	7	6
Milk products	3	2
Meat and alternatives	2	2

SOURCE: Murphy and Barr (2007).

Conclusion

In conclusion, the core concepts that underpin the framework are sound. However, the two-by-two table (Figure 4-1) is an oversimplification of what is needed to apply those core concepts appropriately, and its interface between individuals and populations must be evolved further. We need to separate the needs of the dietetic profession and others who engage in nutritional counseling from applications of the DRIs for public policy purposes (e.g., food guides, fortification, labeling). Furthermore, better supporting documents for nutrition professionals must be developed.

In the end, all applications of the DRIs have the individual as their end user—because people eat food. The question of whether the application relates to individuals or populations is not useful. We must, as we move forward, think about the applications of the concepts, not the numbers. That is the next wave of activity and of thinking in terms of applications of the DRIs.

DISCUSSION

Discussant: Gerard Dallal

The session moderator, Ms. Bush, introduced the discussant and invited him to offer an opening remark.

Discussant Opening Remarks

Dr. Dallal noted that debates about populations and individuals are not useful. From a statistician's point of view, the focus should be on the question to be answered, how the question is approached, the techniques to use, the assumptions to be made, and whether the assumptions and methods are reasonable for the problem at hand. Dr. Dallal also suggested that distinguishing between individuals and populations was primarily an argument about terminology that in the end would not be useful and only distracted from the underlying problem.

General Discussion

Two-by-Two Table

A participant commented that the two-by-two table (Figure 4-1), while useful conceptually, creates problems for users. Policy issues often do not fit neatly into just one box, and some are appropriately placed in all four boxes. The relevant task is to make useful decisions with clear documentation concerning the use of EARs, RDAs, and ULs. In this case, the useful "tools" are efforts to understand the science and the public health ramifications while also incorporating the conceptual and statistical underpinnings relevant to the needed application. Furthermore, she remarked that the decisions are frequently case specific. Thus, by making the primary tool (i.e., the two-by-two table) too simplistic, one loses the ability to be flexible in a rational way.

Another participant noted the use of the two-by-two table as a basic organizing structure and suggested it should not be abandoned. Moreover, she suggested that the use of different DRI values for different applications aligns with the concept of intent-to-treat versus treated-as-intended. When one is working with an individual, as many dietitians do, the intent-to-treat and the treated-as-intended are presumably the same. When one is planning for a large group, such as a food stamp program or school lunch program, the intent-to-treat is not the same as treated-as-intended. This difference explains why one would use a different DRI under different circumstances.

Feedback Loop

Dr. Tarasuk remarked that there are special challenges when a group of experts is asked to address questions raised by government agencies, particularly in terms of clarity about the needs for and ramifications of the outcomes. She expressed interest in ensuring that such efforts include a "feedback loop" to avoid addressing the wrong question or finding that the "real" question had not been specified. She suggested the value of dialogue when these efforts are undertaken to ensure that the potential of the evolving science impacts public policy application.

Development of Dietary Guidance

An audience member asked Dr. Tarasuk about her comparison of the dietary guidance in Canada and the United States. Because the countries have common dietary patterns, how similar or dissimilar were the foods and the end result? Dr. Tarasuk responded that there are several explanations for why the outcomes were similar, despite the use of different approaches. One is that modeling food patterns to achieve the AIs for selected nutrients had a strong impact on the final guidance in both countries. In addition, multiple nutrient goals were translated into food servings for five population groups, so there was a blunting effect for some of the distinctions. Moreover, for both the Canadian and U.S. guides, the nutrient intake achieved by following those guidelines would be far above the RDA for most nutrients. She remarked that this is an interesting example of how a question could be addressed quite differently by two government bodies—depending on whether the perspective was at a public health level or an individual level—and yet the outcome is not very different despite different starting points and assumptions.

Application of the RDAs

An audience member asked about the importance of sensitivity analysis during planning activities, given the uncertainty involved, especially with the coefficient of variation. He also asked how the analysis could be prioritized given the cost and complexity of implementation. Dr. Tarasuk suggested that the use of RDAs for public policy purposes should be re-examined because, in her view, many current applications of the RDAs may not be appropriate; they seem to be carryovers from an older era. Moreover, their use has been driven largely by the question of who the end user is—and the answer is always the individual. However, Dr. Tarasuk noted, this does not mean there is not work that could be carried out for the purpose of counseling individuals, including exploration of the use of ranges or percentages.

PANEL DISCUSSION: IN WHAT WAYS COULD THE GUIDANCE FOR USERS OF DRIs BE ENHANCED?

Panel Members: Danielle Brulé, Mary Frances Picciano, William Rand, and Linda Van Horn

The session moderator, Ms. Bush, introduced the panel members and asked each panelist to offer an opening remark.

Panelist Opening Remarks

Dr. Brulé emphasized the importance of identifying the target audience or users of the DRIs. User needs differ, and the challenge faced by risk managers is to communicate policies and programs to health professionals and consumers who hopefully understand and can appropriately apply DRIs. She further commented that the achievements of the existing framework are impressive and have set the stage for efforts to improve the process, notably documentation and transparency. Dr. Brulé further pointed to the ability of the risk assessment approach to articulate the roles and responsibilities of risk assessors, risk managers, and others. She endorsed a continuing dialogue between risk managers and risk assessors. Finally, she underscored her perspective that ensuring reality checks—for instance, using examples of policies or programs to validate newly revised DRI values—is needed and would be valuable in accomplishing risk communication.

Dr. Picciano expressed a desire to keep the RDAs in the report as goals because social, cultural, or political forces could result in the EARs being misused as goals. There should be more documentation of the assumptions made and their influence on the values derived, particularly for pregnant and lactating women. She highlighted an interest in including discussions in DRI reports that focus on how the EARs and RDAs relate to intake data within subgroups, notably those extrapolated with and without consideration of supplement intake. Finally, she noted that pediatric, geriatric, and reproductive nutritionists should be included in future DRI study committees.

Dr. Rand commented that the interest in "keeping it simple" should be balanced by the value in preserving the inherent complexity. He said it is important to explore the complexities and to avoid plans that ignore these realities or fail to recognize key interactions. Clearly, nutrients have multiple effects, and chronic diseases have multiple risks; such complexities will grow over time. He suggested that when simplification is desired, it would be best to split the tasks more formally so that each has its own focus, including the distinction between science and policy. Simplification could also involve deriving the best estimates of the risk curves together with descriptions of how they have been derived and the assumptions used. In turn, users should provide information on what risk curves are required and, based on their own needs, determine how best to use this information. Finally, Dr. Rand expressed a hope that the future process will include explicit guidelines for risk assessment and SEBR as well as more attention to the "mathematics," such as statistics, modeling, probability, and simulation. In addition, an explicit validation step is desirable.

Dr. Van Horn remarked that ideally the DRIs would be well understood by those who intend to use them. Regarding the role of the registered dietitian and other health-care professionals, she highlighted training and the development of practical tools for implementing and using the DRIs. She also suggested that there were special challenges for practitioners who work within a constant overlay of obesity concerns and must in turn interpret and use the DRIs in the absence of specific interdisciplinary involvement. In terms of reaching out to practitioners and consumers, she stated that, in her own experience with the National Cholesterol Education Program, the "know your number" public education campaign was extremely successful. Perhaps a similar approach could be used for teaching about DRIs. Finally, Dr. Van Horn expressed concern that an exclusive focus on DRIs could detract from efforts to help children select and consume healthful foods and food patterns. She cautioned against the potential for marketing and encouraging the consumption of less nutritionally desirable foods to which nutrients had been added.

General Discussion

Cross-Panel Discussion

Two-by-two table The usefulness of maintaining the two-by-two framework was raised as an issue during the cross-panel discussion. It was pointed out that so many factors are embedded within discussions related to groups versus individuals and planning versus assessment that it is difficult to reflect or "constrain" them within a table. Yet as a broad means of differentiating across the general applications, it is a useful place to begin. Another panel member found it impossible to place her government agency's policies and programs into the four boxes. It was noted that there is a need to have the choice to use the two-by-two table or not use it, as appropriate. Another participant commented that assessment and planning are different ends of the same continuum: If you plan a diet, one of the first things you would do is assess whether it was any good. Another pointed out that the definition of what constitutes a group is also an issue: What are two individuals?

Target population Panel members discussed whether the guidance to users should be enhanced by considering not just healthy populations, but also other populations, such as the institutionalized, the elderly, and "non-healthy" populations. It was noted that there were sections in most DRI reports that focused on special populations and at times also addressed specific disease conditions. However, a further comment was that "healthy" is difficult to define. Moreover, as more than 50 percent of the North American population is overweight or obese, an important question is how obesity fits into these considerations.

Open Discussion

Guidance for use with individuals With respect to the two-by-two table (Figure 4-1), one participant advocated that individuals remain a component of the approach, and that a decision to omit them should occur only after considerable thought and discussion. She emphasized that it was inappropriate to guide individuals based on a population mean. She further suggested that the needed value does not have to be labeled as an RDA, but it should be a requirement plus a safety factor. She added that a reexamination of safety factors to be used may be in order. A panel member posited that the safety factor need not be two SDs as currently used to develop the RDA, but could be one SD or a value anywhere between the EAR and the UL. Another panel member suggested that once a risk curve is established the determination can be made, and remarked that the key point is that the DRI process should specify risk curves.

One person noted there had not been much discussion of "family," as opposed to individuals, small groups, and populations, and that special advice might be needed in this context. In response, it was pointed out that families could be considered small groups of individuals. An audience member indicated her concurrence with the interest in developing more specific guidance for users, but also agreed such guidance should be application specific, not profession specific. She also asked how much time she should spend explaining the current framework to individuals given that changes are likely in the future. A panelist responded that "you have to go with what you have" because we do not know what will happen in the future.

Target population As a follow-up to the discussion about the definition of a healthy individual within the context of DRI development, the comment was offered that the focus on a healthy person may no longer be needed or useful because a large percentage of the population is overweight or obese, hypertensive, or hypercholesterolemic. One participant noted that the focus was on an "apparently healthy" person to reduce confounders, such as those who need blood pressure medication. Another audience member said using healthy persons as study subjects simplifies research protocols.

Consumer messages Dr. Van Horn's analogy to the National Cholesterol Education Program prompted a question on whether a "know your number" target could focus on the concept of a distribution shift. It would be a simple message, but based on distributions and populations; marker nutrients could be used to point persons in the right direction. Another attendee cautioned that the "know your number" concept may work when you have one number, but not when you have multiple numbers. DRIs do not represent all nutrient needs, and pulling out just a few of them has the potential for harm if it becomes a public education effort. People will not turn to food sources of nutrients, but to supplements. Concern was expressed that we need to be careful to target foods rather than nutrients. Dr. Van Horn agreed completely and indicated that her point had been that a well-designed public education campaign worked for raising cholesterol awareness. The lessons learned may be valuable for DRI education and outreach, including awareness of how DRIs can be met by eating the recommended foods.

5

Looking to the Future Process for DRI Development: Session 4[1]

Speakers in the final session were asked to speculate on the future process of Dietary Reference Intake (DRI) development. Specific topics of interest included enhancing transparency of the decision-making process, criteria and "triggers" for updating and reviewing DRIs, determining "new" nutrient substances for DRI development, options for stakeholder input, and important issues that may emerge in the future.

The moderator for the session was Dr. Paul Coates of the National Institutes of Health (NIH). Dr. Catherine Woteki of Mars Inc. opened the session with a presentation on emerging issues and future directions. Dr. Robert Russell of Tufts University then addressed the need to enhance the transparency of the decision-making process. Dr. Linda Meyers of the Institute of Medicine (IOM) gave a brief overview of the options for stakeholder input into the DRI process. Dr. John Suttie of the University of Wisconsin then outlined some types of criteria that could be used to determine when to update or review existing DRIs. Dr. Peter Greenwald of the National Cancer Institute (NCI) examined some considerations in specifying "new" nutrient substances for DRI study.

Discussions open to all audience members were held after each presentation, and a panel discussion in which panel members reflected on what they had heard about DRI development during the workshop closed the session.

---

[1]This chapter is an edited version of remarks presented by Drs. Woteki, Russell, Meyers, Suttie, and Greenwald at the workshop. Discussions are composites of input from various panel members, discussants, presenters, moderators, and audience members.

EMERGING ISSUES: WHAT NEW CHALLENGES MIGHT THE FUTURE HOLD?

Presenter: Catherine Woteki

DRI development is an important scientific undertaking. Future activities can be informed by the conceptual evolution and changes in science that have occurred over nearly 100 years of establishing dietary recommendations for populations. Moreover, it is important to recognize that there is a long tradition of providing guidance on a health promoting diet. Scientists as early as the late 1700s appear to have included concerns about health in advice to the U.S. Congress concerning the creation of the Navy Ration Law. Later, in the 1800s, the U.S. Army Surgeon proposed limiting fat intake and eating a balanced diet; at about the same time the U.S. Department of Agriculture (USDA) highlighted the use of a balanced diet to promote health.

Emerging Issues: 1925–1990

Government recommendations for what constitutes a health-promoting diet can be readily traced to the World War I era, when the British Royal Society produced recommendations for food requirements for populations under stress during wartime. During the period 1925–1937, the Health Organization of the League of Nations began to set estimated requirements for specific vitamins and minerals. In 1933, dietary standards for food programs were proposed by both the British Medical Association and the USDA. An enormous conceptual evolution occurred from 1917 to 1937, with dietary recommendations changing from a focus on food for starvation relief to standards for programs to maintain and improve the health of the population.

A second conceptual evolution was the move from basing recommendations on observations of usual food intake to using the emerging scientific knowledge of the needs for essential nutrients and energy to provide specific nutrient recommendations. From 1941—when the Roosevelt administration asked the National Research Council (NRC) to recommend levels of intake of essential nutrients to maintain the health of the population during wartime—until 1989, 10 editions of the values (called Recommended Dietary Allowances [RDAs]) were issued.

Although the focus was on essential nutrients and establishing estimates of intake to maintain health, concepts of disease prevention were addressed in editions of the RDAs as early as 1958 (NRC, 1958) when a relationship between dietary fat and coronary artery disease mortality was noted. The 1964 edition (NRC, 1964) recommended that adults moderately reduce total fat intake and substitute polyunsaturated fatty acids for saturated fats. The 1968 edition (NRC, 1968) advocated for higher levels of physical activity to reduce the risk of arterial disease, obesity, and diabetes. The 1974 and 1980 editions (NRC, 1974, 1980) contained specific recommendations to decrease calories from fat to less than 35 percent, to decrease saturated fat to less than 10 percent, and to increase polyunsaturated fat intake. The 1989 edition (NRC, 1989a) recommended that the public and professionals look to the IOM report *Diet and Health: Implications for Reducing Chronic Disease Risk* (NRC, 1989b) for more information about dietary intakes as they relate to chronic disease prevention.

Emerging Issues: 1990–2004

As the IOM was embarking on the DRI process in 1994, several issues had emerged that framed the thinking. One was chronic disease risk reduction. Several influential reports had been published, including *Diet and Health: Implications for Reducing Chronic Disease Risk* (NRC, 1989b) and *The Surgeon General's Report on Nutrition and Health* (Office of the Surgeon General, 1988), and the idea that nutrition can play a role in reducing the risk of chronic disease was increasingly recognized. Interest grew in providing a quantitative basis for dietary guidance.

A second issue was the concept of a safe range of intake. There was growing concern about voluntary fortification in the food supply and the increasing use of dietary supplements. It was recognized that there would be considerable value in establishing upper levels of intake and therefore, in a sense, establishing a range of intake conducive to good health. This issue had precedence. Some discussion about nutrient toxicity concerns can be found in the text of early recommended intakes from the NRC, dating back to the 1950s. In the late 1960s, the NRC together with the Council on Foods of the American Medical Association published a policy statement that included the need to set food fortification limits and supplement levels below harmful levels.

New approaches also were considered. The United Kingdom's Committee on Medical Aspects of Food Policy had established multiple reference points in its report on dietary reference values. Dr. Beaton and others had done pioneering work on developing the probability approach. As the DRIs were being developed, new methods for informing public health policy decisions emerged. They included systematic evidence-based reviews (SEBRs), which arose in clinical practice to address the types of evidence appropriate for patient care recommendations; and quantitative risk assessment, which had emerged earlier to assess the risk of non-nutrient substances. Also during that period, results from clinical trials using nutrient interventions against chronic disease endpoints suggested that the single nutrient/chronic disease prevention paradigm had largely failed.

The Future

We can anticipate considerable change and new approaches for making public health decisions in general, and specifically for developing DRIs. Some factors important to the DRIs of the future are described briefly below.

Scientific progress will further our approaches, methodologies, and insights:

- Undoubtedly scientific progress will move us from the consideration of single nutrients to patterns of nutrients or food intake, especially for chronic disease endpoints.
- Divorcing chronic disease endpoint considerations from the establishment of future DRIs will not be possible. The question is where it is appropriate to do so and how.
- Scientific progress will also lead us to focus on "new" nutrient substances and on decision making concerning the adequacy of the science base for the purposes of DRI development for these emerging substances.
- Better statistical techniques for extrapolation and scaling will be developed.
- There will be greater insights with respect to physiological, environmental, and genetic factors.

The public health context for setting DRIs will shift:

- Chronic disease prevention was the overwhelming concern, along with nutrient excess, that framed the thinking in the DRI process. Now obesity is the public health concern through which all of our nutrition problems are being viewed.
- Questions about appropriate levels of fortification and supplementation and the role the DRIs play in informing those decisions will continue, as will questions about formulation of special foods for specific age groups.

A major issue for the future will be the DRI development process:

- Process problems that surfaced during the past 10 years include maintaining consistency across study committees as well as the timeliness, transparency, and openness of the process. In the future, revisions organized by specific nutrients or groups of nutrients should allow greater transparency in addition to better risk characterization, consideration of uses, and verification of the "reasonableness" of those estimates as recommendations.

- Guidance is also needed on how to set the DRIs, the role for SEBRs, statistical techniques, or other new methods to maintain consistency and scientific integrity.
- The risk analysis paradigm provides an excellent model for a future DRI development process. However, it puts additional responsibilities on the government sponsors to provide a clear articulation of the uses of the DRIs. It also puts more responsibility on the IOM in terms of communication within the risk analysis paradigm, communication with the sponsors, and communication with the scientific community.

Implications

The issues raised about DRI development offer direction for the kinds of information needed and identify the steps to be taken to improve future DRIs. We have now identified the key issues, and we have the demonstrated need. Clearly, we can move forward, and clearly, the sponsors and the IOM must embrace larger responsibilities. Furthermore, it is hoped that government agencies that conduct research on nutrition will look closely at these recommendations and develop a concerted research program to address the gaps. The process itself can be improved, but better data are needed.

Open Discussion

An audience member said she shared Dr. Woteki's hopes about the research agenda, but was not sure how it could be organized and who would lead. Dr. Woteki responded by pointing to the need for a broad partnership. She suggested that the starting point would be a rough prioritization of critical research needs and indicated that professional societies can play a key role in this activity. Just as importantly, the political will must be found to ensure funding and related support. In short, the approach must be a concerted effort from a scientific and health perspective.

Dr. Woteki agreed with one commenter who expressed the opinion that the burgeoning interest in functional foods—and, in turn, claims about these ingredients that communicate a sense of requirement—suggest a need for some parallelism between these emerging issues and the DRI development process.

Another audience member noted a recurring theme concerning the need for government sponsors of the DRI process to articulate their uses for

DRIs. She expressed the opinion that this could lead to policy influencing the science and suggested that the DRI study committees should operate independently from issues surrounding uses of DRIs. Dr. Woteki explained that the risk analysis framework she described allowed the independent and unbiased review of the science, but that it also articulated a clear role for sponsors of the assessment to specify the nature of the problem that the assessment is intended to address. Unless those reviewing the science understand why they have been requested to carry out their activities, their outcomes will not be useful. She pointed out that government representatives have expressed concern in the past about the unresponsiveness of the DRI outcomes to their needs and are increasingly recognizing that this may be a function of assigning relatively vague tasks to study committees. Dr. Woteki suggested that the decisions about the nature of questions to be directed to study committees are an enormous and complex responsibility that, under the risk analysis paradigm, includes input from sponsors of the DRI process.

IS THERE A NEED TO ENHANCE TRANSPARENCY OF THE DECISION-MAKING PROCESS?

Presenter: Robert M. Russell

The interest in enhancing the transparency of the scientific decision-making process is by no means unique to the development of DRI values. Others have noted that, as a general matter, a lack of transparency can result in

- perceived inconsistency (whether or not it exists);
- perceived lack of objectivity (e.g., prejudices of study committee members);
- complexity in presentation (rambling narratives that do not provide much information);
- a lack of clarity;
- difficulty in implementation;
- decreased chances of replicability (from one study committee to another); and
- hidden research gaps (Garza and Pelletier, 2007).

Our 10 years of experience indicate there are several points in the DRI process where efforts to make the decisions more transparent would have helped with the clarity of the outcomes and with more ready acceptance of the inevitable scientific judgments needed. Moreover, more information about the decision-making process would have mitigated concerns that study committees did not consider or review certain data or options. In fact, these may have been well considered, but there was no related discussion in the text or "transparency trail" to make this known.

Examples of Transparency Issues

Criteria for Literature Inclusion/Exclusion

One transparency concern that requires attention is the lack of complete "up-front" documentation on the criteria used for literature searches (e.g., what literature was excluded or included). The volume of literature searched was often not explained well in the narratives.

Criteria for Evaluating Evidence

The criteria for evaluating and weighing evidence were also not well documented. Three brief examples—vitamin B_6, zinc, and β-carotene—illustrate this issue. Panels in Britain, North America, and the European Union (EU), looking at the same databases and using the same framework, came up with quite different conclusions. The reasons for this are unclear because the narratives in their reports are not explicit.

For vitamin B_6, in North America, the tolerable upper intake level (UL) was based on a study by Bernstein and Lobitz (1988) in which people taking large amounts of pyridoxine exhibited neurological abnormalities at over 200 mg/day. A no-observed-adverse-effect level (NOAEL) of 200 mg/day was chosen and divided by an uncertainty factor of 2 based on a small number of data on doses less than 200 mg/day. The UL was therefore 100 mg/day. In Britain, they used a study on dogs (Philips et al., 1978) in which the dogs developed ataxia at 3,000 mg/day; that was taken as a lowest-observed-adverse-effect level (LOAEL) and divided by an uncertainty factor of 300 (because of the LOAEL and interspecies and interindividual variation) to derive a UL of 10 mg/day. The EU UL was based on a study by Dalton and Dalton (1987) on self-reported neurological symptoms; the LOAEL of 100 mg/day was divided by an uncertainty factor of 4, based on deficiencies of the database, to give a UL of 25 mg/day. Therefore, ULs of 100, 10, and 25 mg/day were derived by three panels looking at the same data.

For zinc, in North America, the UL was based on a study by Yadrick et al. (1989), which showed a decrease in erythrocyte superoxide dismutase at 60 mg of total intake (diet plus supplementary), to derive a UL of 40 mg/day using an uncertainty factor of 1.5 to account for interindividual variation and use of a LOAEL. In Britain, using the same study, they came up with a LOAEL of 50 mg instead of 60 mg and derived a UL of 25 mg/

day using an uncertainty factor of 2 for use of a LOAEL. The EU rejected that study and instead used balance studies (Davis et al., 2000; Milne et al., 2001), deriving a UL of 25 mg/day.

For β-carotene, in North America, no UL was established because there was no dose–response and the study committee believed the data were conflicting (with no clear explanation). In Britain, they based the UL of 7 mg/day on the Alpha-Tocopherol Beta-Carotene (ATBC) study (and backed this number up by a study in ferrets) in which the equivalent LOAEL was 20 mg/day. In the EU, no UL was established because there were no dose–response data and because different formulations were used in the various studies.

Another example is vitamin A, for which the Food and Agriculture Organization of the United Nations/World Health Organization Recommended Nutrient Intake differs from the IOM RDA (500 μg versus 700 μg for 19- to 50-year-old females). The narratives do not explain why the recommendations differ.

These problems may be solved if SEBRs are used to assist in deriving the endpoints for Estimated Average Requirements (EARs) and ULs. The reviews document the scientific evidence or the evaluation, document and rank the uncertainties around the estimations, and rank the health implications of the intakes above and below the reference intake. There is provision of a rationale, in the form of extensive tables, for decisions reached, replacing the rambling and vague narratives in some reports.

Subjective Uncertainty Factors

The factors used to account for various sources of uncertainty were highly subjective and varied depending on the study committee. Bias can be minimized, however, by following predefined rules—not by using an uncertainty factor that aims for a convenient UL or a UL above the recommended intake, for example.

Endpoint Selection

More transparency would have helped with the concerns that surround the selection of endpoints used. Although there is a presumption that these choices were made for public health protection and significance, the decision process is not clear from the narratives.

Specification of Research Gaps

An important activity carried out without benefit of an identifiable and accountable set of criteria was the specification of research gaps. Apparently the research gaps identified by study committees were often compiled at the last minute, after the committees had completed their exhaustive work of deriving EARs and ULs. In short, research gaps were developed too rapidly and without enough deliberation. An overall effort at transparency would have called attention to this approach and perhaps improved it, making it more deliberative.

Configuration of Study Committees

Finally, improved transparency has the potential to ameliorate the nearly inevitable problems associated with configuring study committees. The configuration of study committees may introduce bias, particularly when the committees (which were limited in size) had to deal with large numbers of nutrients. For example, the micronutrient committee had 14 members to study 14 nutrients. For a number of nutrients, there was only one expert on the committee. It is possible that a single person's opinion could go unchecked, particularly if he or she had an assertive personality. Transparency and accountable documentation for the decision made could help in terms of double-checking outcomes and explaining the reasons behind decisions.

Implications

The DRI development process would be improved, and transparency enhanced, if more active and targeted efforts were made to explain and document the decisions made. More specifically, SEBRs should be used to help make key decisions on endpoint selection, for both EARs and ULs, with thorough documentation tables. Rating scales for uncertainty and public health importance should be used as an important part of the evidence-based review process. Also, predefined rules should be followed for uncertainty factors to minimize bias.

Although transparency can help to mitigate the effect of one strong member of a study committee when it may have only one expert on a certain topic, a better solution might be to focus on a smaller subset of nutrients—a single nutrient or small groups of nutrients that interact (e.g., vitamin D, calcium, phosphorus, the antioxidant nutrients; folate, vitamin B_{12}, vitamin B_6, riboflavin; sodium, potassium)—so that more depth in expertise can be available on the committee and knowledgeable group discussions can be used to reach conclusions rather than the perspective of a single scientist.

Open Discussion

A participant expressed doubt that SEBRs would be helpful to DRI development. He questioned how scientists who are not nutritionists could judge the relative importance of studies germane for setting reference values. Dr. Russell replied that SEBRs could enhance some critical components of the development process, but do not replace activities that would focus on scientific judgment and decisions made by those with nutritional or other appropriate expertise. Furthermore, such reviews would not be needed—nor are they necessarily appropriate—for every decision, every nutrient, or every DRI value.

Another participant suggested the drive for consensus in the DRI process may have obscured important disagreements. She asked whether enhancing transparency, as described by Dr. Russell, would avoid this, and indicated it may be better to explicitly state and explain the disagreement rather than forge a consensus. Dr. Russell responded that an SEBR would have helped the vitamin D discussions, for example, with regard to the appropriateness of the 25-hydroxyvitamin D level as an indicator. He added that there is considerable value in making available an objective scientific evaluation of the data via an SEBR rather than depending on the potentially biased opinions of strong personalities.

An audience member commented that the inability to openly discuss challenges is problematic and could limit transparency. It was suggested that there could be points in the process when outside advice should be sought. Dr. Russell agreed. In the future a feedback loop may be possible, as long as there is no possibility that stakeholders could influence the science or compromise the scientific independence of the study committee.

One participant noted that, with respect to the β-carotene example highlighted by Dr. Russell, a clear problem formulation step or an indication of the uses of the needed reference values, particularly regarding the intended population, would have helped to enhance clarity. She then asked about the composition and functioning of study committees, suggesting there were two models. In the first model, study committee members are likely to have been investigators for some of the studies that will be reviewed and used in deriving the DRIs. In the second—a consensus model approach, as is common at institutions such as NIH—the study committee members have relevant expertise, but their studies will not be under review in terms of the scientific evaluation. She asked which would be more helpful for transparency and scientific rigor. Dr. Russell responded that study committees should have a meaningful number of people who are specific experts, and expressed concern that a consensus-type approach would not solve all the problems associated with study committee bias, adding that more than two members on a study committee per nutrient topic area seemed to be another important avenue to pursue.

OPPORTUNITIES FOR STAKEHOLDER INPUT IN IOM ACTIVITIES

Presenter: Linda D. Meyers

This presentation focuses on the "rules" surrounding study report development at the IOM and the options for input. To address these topics, we need to understand what the IOM is, why advice is sought, how IOM functions, and what motivates the procedures for input.

Role of the Institute of Medicine

The IOM is generally regarded as a trusted independent advisor. The government approaches the IOM for consensus advice that is not influenced by any particular group (including the government) and that draws on the best minds in the country. The IOM was established in 1970 under the charter of the National Academy of Sciences. It is part of the National Academies complex that encompasses three honorific societies: National Academy of Sciences, National Academy of Engineering, and Institute of Medicine.

The IOM is a private, independent, nonprofit, "soft money" organization, with no line appropriation from Congress. It has 1,600 elected members and about 130 staff in 6 program units, one of which is the Food and Nutrition Board (FNB). It publishes about 40 reports a year.

The IOM—and the entire National Academies complex—engages in activities that include committee studies, workshops, forums and roundtables, symposiums and lectures, expert meetings, and communication functions. Communication activities include so-called "derivative products," which are based on existing reports and contain no new recommendations.

Nature of Consensus Report Development and Opportunities for Input

Reports produced by consensus committees tackle major health issues, such as DRIs, obesity prevention, overhaul of the Food and Drug Administration's drug review system, medical errors, and health literacy. As the backbone of the Academies' activities, the committee studies are the most visible. They are prepared by balanced expert committees. Experts serve without remuneration, an effort designed to increase their independence. The committees strive for full consensus based on evidence, and their work is evaluated through a rigorous peer review process.

Many current policies for the process of report development derive from Section 15 of the Federal Advisory Committee Act (FACA). Section 15 was passed in 1997 and is intended to maintain the Academies' independence from government. It requires the Academies to ensure public input into committee activities and provides opportunities for transparency. It is part of the standard procedures under which the IOM now operates.

As a result of Section 15, the committee appointment process includes posting biographical sketches; data-gathering meetings are open to all; closed-meeting summaries are posted; and there is public release of reports delivered to sponsors (unless a report is classified). If the committee process has followed FACA Section 15, U.S. federal agencies may use the committee's advice or recommendations.

The consensus study process provides several opportunities for input, as shown in Figure 5-1. One is the definition of the study, called problem formulation in the risk assessment framework. This is a time for working closely, usually with the sponsor, to define the task statement.

The second opportunity is the committee selection and approval process. During this phase, nominations for individuals to serve on the committee are sought. Once a provisional committee is appointed, committee biographical sketches are posted on the National Academies Current Projects website (http://www8.nationalacademies.org/cp/) and there is an additional opportunity for comment.

The next phase, committee meetings, includes open and closed sessions. Open sessions are open to all and include the gathering of scientific information, discussions with sponsors to ensure understanding of the statement of task, and workshops. The committee deliberates in closed sessions to allow its members to debate ideas without fear of outside influence and to change their minds as they consider evidence, which is all part of the process. Written materials given to the committee from the outside are put in

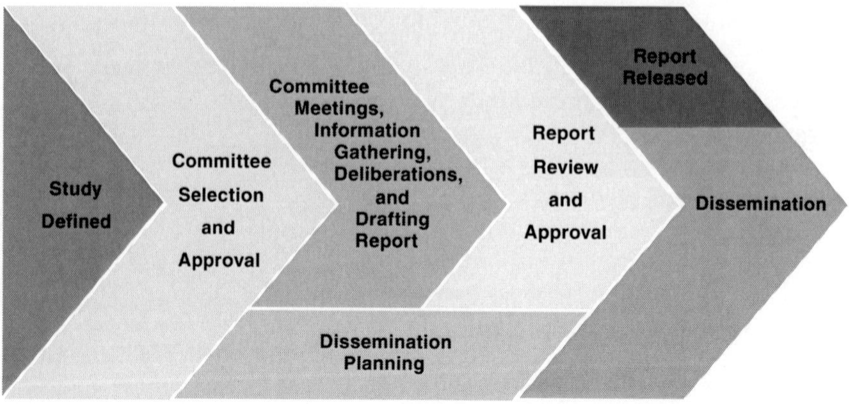

FIGURE 5-1 Opportunities for input into the consensus study process.

a public access file for the benefit of the public and transparency, although it is difficult to obtain proprietary information under these conditions.

The report review process is closed, and even the reviewers are asked to keep the reviews confidential, again to allow them to freely express their opinions. Report release is public, and dissemination and dissemination planning are activities that are intended to be done collaboratively and with a lot of input.

Opportunities for Input Through Workshops and Forums

Workshops like this one provide another opportunity for input. Workshops are usually public. They may be part of a study or stand alone. They often result in a meeting summary, have a formally appointed planning committee, and are intended to raise issues and discuss and hear individual suggestions. This workshop also allowed the opportunity to comment on background papers.

Forums—also called roundtables—have become a powerful way to energize the field and develop new ideas and insights that could be used to spin off studies. They also serve to engage a broader range of the scientific community, and they provide a bridge for communication, often among academia, industry, government, and consumers. Forums and roundtables are set up intentionally to have a range of perspectives. The members set the agenda. Forums and roundtables may also commission background papers and sponsor workshops. There are no consensus recommendations, but there can be ideas that lead to consensus studies. The 12 forums and roundtables currently in the IOM range from drug discovery to evidence-based medicine to food to neuroscience and nervous system disorders.

Summary

The IOM was established by government to advise government while being independent. Sponsors of IOM activities come to the IOM for its credibility and independence, so protecting and achieving those two elements are a critical part of the Institute's procedures and mode of operations. FACA Section 15 is also important to achieving credibility and independence. Input can be provided through a variety of mechanisms, and within the procedures, there is room for creativity and flexibility.

Open Discussion

An audience member suggested another option in addition to the use of IOM study committees as the source of scientific reviews to inform DRI development. That option is to have branches of government appoint advisory committees, as is done for the dietary guidelines in the United States. She asked for Dr. Meyers' views of the advantages and disadvantages of the IOM process compared with an advisory committee approach. Dr. Meyers noted that the IOM ensures scientific integrity by providing independence—a "closed door" when appropriate—and the ability to attract high-level scientists in a range of disciplines. Her personal experience in government suggested that when the *Dietary Guidelines for Americans* committees had access to science reviews prepared by the IOM or in some cases NIH, the model for the development of guidance worked well. However, when this scientific input was lacking, the government advisory committees for the dietary guidelines experienced many more challenges.

Another participant added while transparency and openness in government activities are desirable and should be respected, there is a certain amount of disingenuousness about accomplishing the needed tasks while being so fully open. She posited that these government committees actually "meet in their hotel rooms at night" for the needed discussions because the opportunity for a free give-and-take discussion of the issues is inhibited by the public nature of the sessions. Moreover, she suggested there is no disconnect between the process used by the IOM and the open process required by the government. She suggested that the IOM has been highly effective and appropriately responsive to FACA, ensuring flexibility and openness when needed while specifying closed sessions for certain aspects of deliberations.

One commenter suggested that the current structure for DRI development has in the past failed to enable the process to readily include government users of the DRIs, especially within working groups and as part of the early deliberations of DRI development. He advocated creative solutions to bring in the expertise that is clearly available, but inadequately tapped. Dr. Meyers agreed that this was desirable.

WHAT ARE THE CRITERIA OPTIONS FOR DETERMINING WHEN TO UPDATE/REVIEW EXISTING DRIs?

Presenter: John Suttie

Given the prospect that DRI development in the future will not involve routine updates of a large group of nutrients, as has been done in the past, considerations about the criteria and process for updating the DRIs take on a great deal of importance. The factors that might "trigger" the need to update existing DRI values can take several forms. Moreover, the way in which these triggers can be identified, considered, and acted upon requires a previously agreed-upon process along with an established set of criteria.

Initiating the DRI update/review process has at least three possibilities. Decisions to update could be generated by the IOM, which would be responsible for determining that a specific nutrient should be revisited and seek funding from potential sponsors. A second possibility is that stakeholders could generate the request. Under this scenario, stakeholders would likely contact both the IOM and potential sponsors to generate the needed activities. The third possibility, although this has not occurred, is a congressional mandate.

More specifically, the process could be driven in several ways. Approval of a petition from stakeholders would be one mechanism. Another approach would be the use of a specifically established IOM standing committee. Assuming that criteria were set in advance, such a committee could examine research advances, then refer them to the IOM for possible action. It could be tasked with a focus on advances in research regarding the biological knowledge about the nutrient(s) and/or focus on the needs of the user community, suggesting that a problem or issue needs to be reconsidered.

From the perspective of one who has had many years of experience with the DRI process, a few additional comments can be made. First, as many users seem to find that the AIs present challenges, perhaps nutrients with AIs need to be revisited in order to derive values that can be better used, such as EARs/RDAs. Second, the apparent need to review nutrients on a single-nutrient basis rather than reviewing all nutrients over a 5- or 10-year period has some advantages. We will have the opportunity to put more experts on the study committee who are familiar with the specific nutrients and their research areas, as well as additional specialists, such as biostatisticians, who could not be included on previous study committees. This will facilitate more fruitful deliberations. Also, given the likelihood of more targeted case-by-case updates, it would be wise to post the previous edition electronically and then update as new data become available. That is, the new reports should be published in an electronic "loose-leaf" format.

Finally, prioritizing the need for revised DRIs is an important consideration. New and relevant data are an important threshold consideration. Also, the quality and number of new data available would be expected to impact the prioritization. As many have already pointed out, an important factor for prioritization is the public health significance of the nutrient.

Overall, it seems fair to suggest that these update and revision possibilities can readily be considered and an approach with relevant criteria put in place. What may be more pivotal are the needed commitments: commitments from the IOM to play its needed role and a commitment from potential sponsors to recognize that nutrients may need to be revisited and to see that these efforts are funded.

In sum, we have benefited greatly from working with the DRI sponsors during the past years and especially during the planning process of this workshop. I believe all groups now have a better idea of what is needed and how we have to work together for a useful and appropriate outcome. Our next task is to make this a reality.

Open Discussion

An audience member noted that the IOM had established α-linoleic acid as the "essential" omega-3 fatty acid,[2] rather than docosahexaenoic acid (DHA). In response to the question on whether a nutrient that is physiologically essential could be considered essential to one's diet, Dr. Suttie said this could be examined in the future. Another participant noted that the existing reference values lose credibility as medical societies and other organizations offer different recommended intakes, presumably based on newer data. The concern expressed was that timely updating needs to occur so that DRIs, currently based in some cases on data more than 10 years old, do not become irrelevant. Dr. Suttie agreed there is value in working urgently to provide updates needed. With respect to Dr. Suttie's warning that a congressional mandate was an option for initiating DRI review, one audience member was concerned that Congress may lack the ability to target key questions appropriately and may thereby undermine the entire process.

A question was raised as to whether the updating approach needed to focus on a total revision of all aspects of the reference value, or if targeted or partial revisions—for example, updating values for one age/gender group—would be appropriate. It was also noted that there may be another type of revision that could be categorized as a technical correction, assuming a factual error or similar problem has occurred. Dr. Suttie responded that these were all open options. He suggested that actively working to form a group to deal with these questions was an important first task. He also pointed out that the possibility of a loose-leaf format and the availability of electronic files might allow these small or technical changes to be more easily managed.

WHAT ARE THE CONSIDERATIONS IN SPECIFYING "NEW" NUTRIENT SUBSTANCES FOR DRI STUDY?

Presenter: Peter Greenwald

This presentation addresses the challenges we face in ensuring that

[2]An Adequate Intake (AI) and an Acceptable Macronutrient Distribution Range (AMDR) were developed for α-linoleic acid.

the effects we attribute to nutrient substances relative to health/disease outcomes are real and likely to be stable over time. The needed discussion begins with the broader issue of the use of research to inform public policy, then moves to examples germane to nutrient substances.

Threshold Considerations

The first task related to the use of research to inform public policy is whether there are enough research data to even begin. In the field of nutrition, there are often not enough data because of insufficient investment in the basic or clinical nutritional sciences to provide the groundwork for these considerations. Despite its great importance, biomedical research on nutrition is starved for resources. Clearly this issue needs to be addressed.

A second task is to enlist balanced, expert committees that follow a systematic approach, using explicit criteria to synthesize research results through a transparent process. The IOM often, but not always, does this. With nutrient substances, the evidence usually is complex and conflicting, and sometimes one study will abruptly make all the other data obsolete. A group experienced in such SEBRs is the U.S. Preventive Services Task Force; its approach is described in Box 5-1.

Another resource reflective of vast experience in this area is NCI's Physician Data Query (http://www.cancer.gov/cancertopics/pdq/cancerdatabase). It works to synthesize evidence, then uses the outcomes to inform professionals and the public. It ranks evidence according to various levels. For

BOX 5-1
Approach Used by U.S. Preventive Services Task Force

The Task Force, under the auspices of the Agency for Healthcare Research and Quality, has wide experience in systematic evidence-based reviews. It recommends the following steps for rigorous review of the evidence:

First, create an analytic framework:

- State the key questions, and prioritize them.
- To attract agency interest, request that the discipline define the studies that would give a clear answer to the questions.
- Explicitly state the information that will either confirm or refute the ideas.

Second, systematically review the literature:

- Rate the quality of each study in terms of its ability to answer the key questions.
- Examine the benefits and harms, then determine the balance among them.

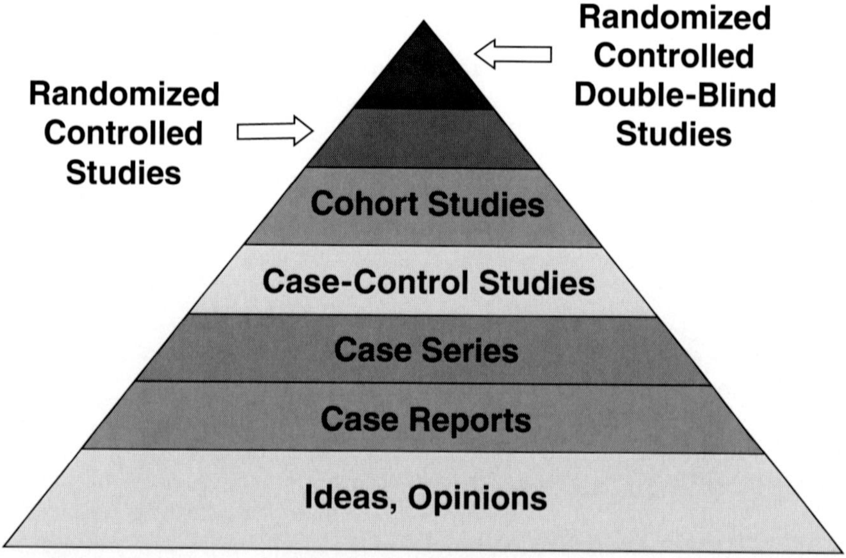

FIGURE 5-2 Amount of research done, by type: The smallest amount of research is conducted using randomized clinical trials (most useful).

our purposes, it is important to note that the most definitive studies are randomized controlled clinical trials (buttressed by basic nutritional science), followed by non-randomized controlled trials. The studies become weaker from that point, with cohort or case-control studies and ecologic studies appearing as one moves down the pyramid. At the very bottom of the list are the opinions of respected authorities.

Figure 5-2, developed by Dr. Barry Kramer, underscores the challenge we face. The pyramid shows the amount of research in each area mentioned above. Little research of the most useful type (randomized clinical trials) is available, whereas there is an enormous amount of information that is not very meaningful. This needs to be reversed.

Nutrient Substances: The Nature of the Evidence

As we learn more about individuals and individual variation in genomics and related fields, we find changing evidence for nutrient–disease relationships. We have to develop an approach to consider the changes people experience over their lifetimes, their different sustainabilities and reactions to foods, and the changing environments to which they are exposed. To illustrate these points, two studies will be discussed.

Linxian Nutrition Intervention Trial

The first study is a 5-year trial started 25 years ago in collaboration with Chinese scientists and conducted in rural China, where the diet is borderline deficient and the population is very stable. The main focus was on esophageal cancer, including cancers of the upper stomach. Relevant study information includes the following:

- *Study population:* Nearly 30,000 adults aged 40–69.
- *Study design:* Factorial for comparison of four groups of nutrients: (1) retinol and zinc; (2) riboflavin and niacin; (3) ascorbic acid and molybdenum; and (4) selenium, β-carotene, and vitamin E (factor D).
- *Outcomes:* Factor D decreased total mortality by 9 percent (Figure 5-3), decreased total cancer mortality by 13 percent, and decreased total gastric cancer mortality by 21 percent (Blot et al., 1993). However, whereas factor D decreased total mortality for those who were under 55 at the start of the trial (1986), the benefit largely occurred after the intervention was stopped (1991). The fact that the time of exposure may not be the same as the time of the benefit must be considered with nutrients. In the people aged 55 and above, it made no difference at all.

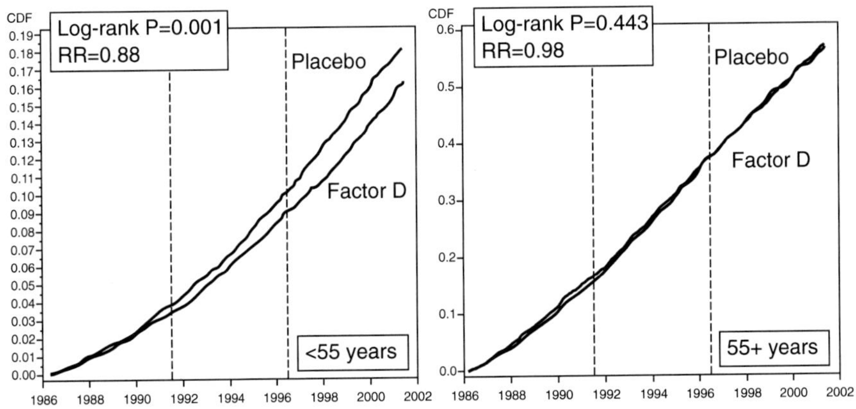

FIGURE 5-3 Linxian Nutrition Intervention Trial: total mortality decreased in <55 years age group by factor D.

Also in this study, a similar pattern was observed for total cancer mortality and stomach cancer mortality (Taylor et al., 2005).[3] This suggests that the intervention had an impact in the precancerous period. Complicating the issue further was the observation that what benefited overall mortality, total cancer mortality, and stomach cancer mortality appeared to be detrimental against esophageal cancer mortality. This indicates that the overall effect must be considered if assessing the effects of nutrients on a population at large or major subgroups of the population.

U.S.–Finland ATBC Lung Cancer Prevention Trial

This was a study using 29,000 men who were heavy, long-term smokers. In the early 1980s, when the trial started, most epidemiologists believed that β-carotene prevented cancer. The trial was factorial. The participants took vitamin E, placebo for vitamin E, β-carotene, and placebo for β-carotene. There was no difference in lung cancer incidence for about 4 years. After that, however, the curves started to separate (Figure 5-4), and smokers on β-carotene began to do worse. In the β-carotene group, the risk of lung cancer was about 6 per 1,000 men per year, whereas in the placebo group, it was about 5 per 1,000 men per year, a 16 percent difference. The important message is that a large, well-designed, well-managed, double-blind clinical trial was needed to detect that 1 per 1,000 difference. Without this trial, people might still believe today that β-carotene protects against cancer.

There was also a one-third reduction of prostate cancer occurrence in this trial in the men taking vitamin E. With about 14,500 men on placebo for vitamin E, 147 got clinical prostate cancer. With 14,500 on vitamin E, 99 got clinical prostate cancer. As further evidence of the value of this type of work, the observation of a secondary endpoint, together with other data, led to the design of a study of 32,000 men now in progress, to examine whether vitamin E and/or selenium will reduce the occurrence rate of clinical prostate cancer.

Implications

Regarding efforts to establish the impacts and risks/benefits of nutrient substances, the following points are worth noting:

- Efficacy data based on sound scientific evidence must be present before making public health recommendations regarding nutrients.

[3]Also personal communication, P. Taylor, National Cancer Institute.

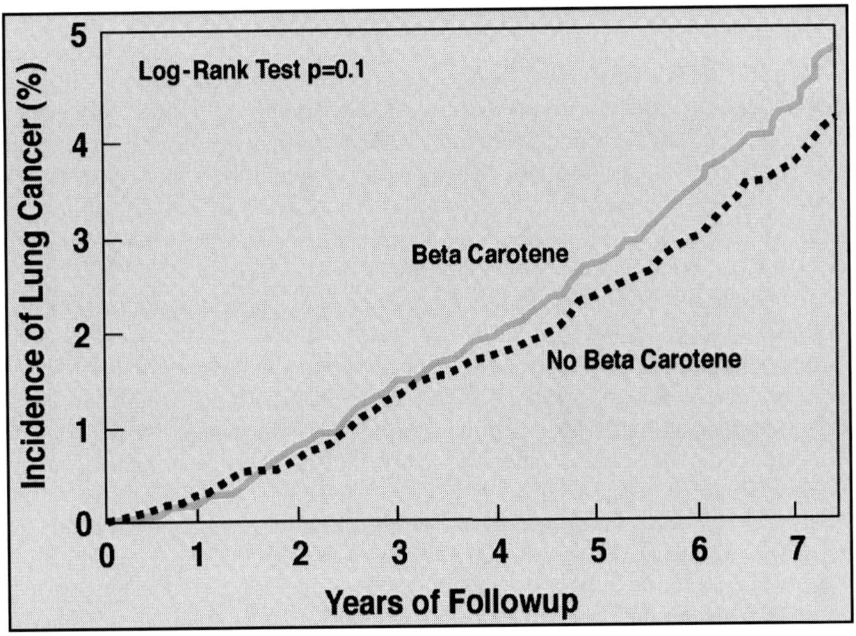

FIGURE 5-4 Cumulative lung cancer incidence in U.S.–Finland ATBC Lung Cancer Prevention Trial shows that beta carotene does not protect against cancer.

- Many existing data are not sufficient, not sound, and even contradictory; these need to be sorted through using systematic approaches.
- Confidence in nutrient–disease relationships can change, often in unexpected directions.
- Large randomized trials have the greatest impact in changing the level of confidence in a nutrient–disease relationship. Although these trials have an enormous cost, they are necessary.
- We need greater investment in research in the nutrition area.

To underscore the importance of "getting the science right," we need only turn to a recent article in the *New York Times Magazine* written by a respected science reporter. It was entitled "Why can't we trust much of what we hear about diet, health and behavior-related diseases?" (Taubes, 2007). The reporter includes several examples, many in nutrition epidemiology, where there is so much conflicting evidence that people do not believe it. Clearly, we have a serious problem, and we must push for the conduct of definitive studies before we make pronouncements on public health.

Open Discussion

One commenter noted that the research pyramid shown by Dr. Greenwald did not contain animal or in vitro studies. Dr. Greenwald responded that although they were not included, animal studies are important if they are relevant to humans. Another participant added that understanding the underlying mechanism is one of the enormous values that comes from animal studies. Dr. Greenwald agreed, provided again that the studies are relevant to humans.

An audience member asked Dr. Greenwald to comment on obtaining funding for research within the nutrition community at large. Dr. Greenwald responded that existing groups within the nutrition field, including the central coordination at NIH, tend not to be aggressive in looking for resources, and this shortcoming needs to be remedied.

A participant emphasized his concern that randomized trials for nutrition lacked intermediary variables that in turn impact the plausibility of the effect. He stressed the desirability of establishing plausibility as it relates to clinical trial design. Dr. Greenwald responded that this was a good point, and that while biomarkers can play an important role, too often measures cited as biomarkers are only reflective of associations. Unless a biomarker is demonstrated to be predictive of the outcome of interest, such measures are not sufficient.

An audience member expressed the opinion that there were functional food factors that were effective in preventing cancer and asked about the desirability of informing the public through food labeling. Dr. Greenwald replied that the data are too sparse to support such activities. He emphasized that a true disease endpoint must be identified, covering multiple categories of disease over a sufficient period of time and taking into account individual variability.

A participant asked about an approach for food components such as macronutrients that cannot easily be studied in randomized clinical trials or similar types of research. Dr. Greenwald responded that the needed investment in basic nutritional science should specifically include efforts to clarify the role of such substances in different cells, organ systems, and models. He further suggested that there may also be a role for clinical metabolic studies to provide initial and clarifying data on the effects in humans. In turn, in each case, there needs to be a debate as to whether there are sufficient data to allow conclusions to be drawn and research to proceed. It would be important to avoid beginning an expensive large-scale, long-term trial without first having the basic information that would ground the rationale for the trial.

PANEL DISCUSSION: REFLECTIONS ON WHAT WE HAVE HEARD ABOUT THE PROCESS OF DRI DEVELOPMENT

Panel Members: Mary Bush, Jean-Pierre Habicht, Suzanne Harris, Van Hubbard, and Molly Kretsch (later joined by Stanford Miller)

The session moderator, Dr. Paul Coates, introduced the panel members and began the discussion by asking each panelist to offer an opening remark.

Panelist Opening Remarks

Ms. Bush expressed her belief that the experience of the past 10 years has greatly advanced the DRI development process. The fundamental task will be to ensure that intellectual rigor, as offered by the IOM process, is applied to decision making throughout the next steps for DRI development. She added that timing and resources will be critical considerations. Ms. Bush emphasized the importance of communication, including dialogue at key points between sponsoring agencies and the IOM as well as communication with stakeholders regarding policies and programs based on the DRIs. It was noted that the way in which the scientific concepts are "put together" in the DRI reports is essential to stakeholder understanding—and acceptance—of the process and the reference values and the process used to create them. The process must be clear, transparent, and understood by stakeholders, or the credibility will suffer.

Dr. Habicht commented that the scientific work to develop DRIs in a fashion that permits them to be used effectively is still not well advanced. The first task is to deal with the lack of attention to the uses of DRIs. The purpose of DRIs must be clearly enunciated and the objectives of the work related to this purpose. He suggested that consideration of uses earlier in the workshop could have resulted in discussions helpful in remedying the current shortcomings of the DRIs in terms of their applications. His interest was in identifying what is necessary and in categorizing applications because approaches for guidance are developed application by application. For example, despite the attention given to the need for information on the distributions of requirements, for many important applications of the DRIs there is no need for information about the requirement distributions. A second example relates to dietary counseling of individuals for whom the dietary guidelines may be the best first-line tool in nutrition counseling; only when this tool reveals special problems does one need to turn to DRIs. In short, the appropriate sequencing of tools has not been systematically examined. For this reason a discussion of timing of updates and prioritization of nutrients is premature until systematic, scientific investigations based on considerations of use have been undertaken. He also suggested examining the implications of the imprecision of the coefficient of variation for various applications as well as the need to reexamine the "false" AIs that exist for some nutrients.

Dr. Harris pointed out that there is still much to do to make DRI outcomes and tools as useful as possible for stakeholders. Much can be gained by identifying groups that are interested in such work and willing to take on collaborative efforts, and this would be assisted by a facilitating or organizing mechanism to track the activities. For instance, the efforts being carried out in Europe can be informed by the DRI work, and DRI development can benefit from their input as well. She noted several key topics discussed during the workshop, including separating the DRIs from the dietary guidelines, adding nonessential nutrients important to health to the list of substances considered, and the merits of updating reference values on the basis of single versus groups of similar nutrients. She also noted value in looking for tools to improve the consistency of UL development. Dr. Harris supported the benefits of risk assessment and its discipline of decision making. Finally, she suggested that changes in the food supply should be taken into consideration and we should ensure the availability of high-quality food composition databases as well as relevant expertise within DRI study committees.

Dr. Hubbard stated that the onus is on all of us to be more specific about what we are defining, and then to work to understand how it can be applied best to individuals, groups, special conditions or diseases, and agency/government planning or policy activities. He compared the process and its principles to the evolution of transportation, which changed in response to both scientific/technical development and needs/purposes. From his perspective, key recognitions that must be embraced include the following: values are best estimates and it is appropriate to assign some probability or level of uncertainty to them as we apply them to individuals or groups; interactions among nutrients influence their functionality; metabolic changes within the body may occur as we increase the intake of certain nutrients, and this may alter requirements for other nutrients; and the continued focus on chronic disease reduction reduces the specificity of the biomarkers used to set recommendations. As a final point, he suggested that we need to carefully consider the modifications in the process that may be required if we decide to base DRIs on the total population rather than "healthy" persons.

Dr. Kretsch addressed the need for research. She highlighted the many research gaps, including endpoints, biomarkers, life stage data, testing the recommendations, and dose–response data near the EAR. She acknowledged that although new research is needed, there is no certainty that research will be conducted. She expressed particular concern about the current opinion that basic research in support of the DRIs is not innovative and will not attract the interest of young scientists. She called for a change in this attitude and stressed the importance of basic nutrition research to reduce uncertainty surrounding the DRIs. She closed her remarks by suggesting it would be useful to review issues related to DRI values and the ability of the population to meet them. Such concerns are very real for practitioners and policy makers alike and warrant attention in the future.

General Discussion

Dr. Coates opened the discussion to all attendees. However, he first asked panelists if they had additional comments.

Guidance for Users

One panel member asked whether, as suggested by a presenter in an earlier session, bringing together a group to focus on a particular application to derive more relevant or specific guidance might be useful. Another responded that we do need to consider the guidance application by application to improve the guidance offered. A comment was added that the definition of "small groups" is problematic; the two-by-two table specifies individuals and populations, but smaller groups are not addressed.

An audience member remarked on the complexity of improving the eating patterns of North Americans given the need to synthesize a great deal of information coupled with the diversity of the food supply. A panel member agreed that it is a challenge to use all of the information to create meaningful dietary patterns, but that it is a needed task so that the guidance given does not require consumers to deal with technical details.

Research Priorities

A question was asked about approaches for setting priorities given that the study committees had identified research gaps. A panelist suggested it would be useful to identify a coalition of stakeholders who would specify the top five research priorities based on the existing IOM effort to synthesize the research gaps. An audience member suggested that participants should sponsor junior colleagues to attend future events such as this workshop in order to foster interest in relevant research.

Food Composition Databases

An audience member expressed concern about some of the shortcomings in data that underpin existing tables of food composition. The impact of digestibility and the impact of processing were two examples given. In response, it was noted that for some nutrients the absorbability in the usual North American diet was specified in such tables. One commenter then pointed to the critical need for food databases to keep pace with the food supply so that the proper context for DRI development is available.

Funding from For-Profit Groups

A question was directed to Dr. Meyers concerning financial support for the DRI project, and whether there was an effort to restrain outside funding from for-profit groups to a level lower than what the Academies allowed. Dr. Meyers indicated that the general policy for the National Academies is that no more than 49 percent of a project's funds should derive from for-profit entities, to ensure independence, but that these percentages are determined on a case-by-case basis for each project, and the percentage is generally lower. For the DRI project, a small percentage of support was obtained from the Dietary Reference Intakes Private Foundation Fund and the Dietary Reference Intakes Corporate Donors' Fund.

International Collaboration

An audience member noted that the European Food Safety Authority (EFSA) was tasked with producing population reference intakes for nutrient substances, including macronutrients. Its report on macronutrients should be available for public consultation next year. She suggested that IOM input during this consultation would be greatly appreciated. EFSA also has a second task—to advise the European Commission on how to turn nutrient reference values into food-based dietary guidelines. She contrasted this approach to that used in the United States.

A participant then emphasized that international collaboration would provide opportunities to learn from each other and to conserve resources. She asserted that it would be extremely valuable to have at least an informal collaboration to create awareness. Another audience member expressed interest in time lines for next steps given the activities in Europe that would benefit from the DRI work. In response, it was noted that some issues can be resolved more easily than others, but the tasks have to be done correctly and may take time.

Derivation of AIs

A participant asked Dr. Habicht to clarify his statement on the need to set priorities to reexamine "false" AIs. Dr. Habicht responded that "false" AIs appear to DRI users to have been based on the well-established approach of deriving a recommended intake using a population intake level. He stated that "false" AIs are based on something else. A member of the audience indicated that one approach to developing an AI is to use the median intake of a healthy population, but that AIs had also been developed using an adequate intake level from an experimental study group, and these values are perhaps the ones Dr. Habicht considered to be "false" AIs. Dr. Habicht recommended that the term "AI" be retained for "what it really is" and that another term be used for other approaches.

Special Topics of Interest

An audience member made a plea to revisit the AI for fluoride, noting the disparity between the AIs for children 0 through 6 months and 7 through 12 months. The concern was that the AI for the 0- through 6-month age group was too low to afford the needed public health protection regarding the development of dental caries later in life. Another audience member asked about the practicality of making changes and the possibility of changing values. The response was that if a value cannot be supported, it should be removed or changed.

Another participant addressed the question of "other food components," such as carotenoids, β-carotene, and flavenoids, or what are referred to as *bioactives*. He suggested that they currently appear to fall outside the DRI process, but he believed that the task of evaluating the science and making relevant recommendations for such substances fell within the purview of the IOM.

At the end of the discussion, Dr. Coates invited **Dr. Sanford Miller** to join the panel. He noted Dr. Miller's considerable experience with DRI development. Dr. Miller offered several comments. He noted that AIs were developed primarily as "placeholders" because no other data were available to allow a recommendation to be made. This approach was taken because failing to issue a reference value fosters the incorrect conclusion that the substance is safe at any level, a conclusion DRI developers wished to preclude. Additionally, he pointed out that the controversy about AIs is broader than it appears. Specifically, he identified the broader issues as follows: What do you do when you lack data? What is the default to be used? Can the data be developed, and how do you present the data given the uncertainty surrounding them? Dr. Miller also focused on the question of precision. He indicated that accuracy is needed, but asked how much precision is needed, suggesting an inordinate number of decimal places were used during the development of DRI requirements.

Dr. Miller expressed his concern about the possibility of marginalizing the research that should be at the beginning steps of DRI activities. Although the research needed is often identified as clinical in nature, Dr. Miller countered that the importance of basic *biological* research cannot be overstated. Such studies outline the physiology involved, and offer a direction for focused human research. He underscored that animal data are not used to extrapolate values for humans, but rather provide information for designing the needed human clinical trials.

Finally, Dr. Miller remarked on the considerable number of crosscutting issues, the advances that have been made, and the back-and-forth between seeking the advances and returning to basics. He highlighted the importance of "finally beginning to understand the process" and closed by remarking that 14 years ago, this meeting could not have taken place—it is not that the questions have changed, but how we are asking the questions has changed.

6

Summary and Closing Remarks

Presenter: John Suttie

Clearly the Dietary Reference Intakes (DRIs) are important and widely used. The process for their development is a critical activity worthy of the serious consideration given by this audience. I would like to thank all of you for your thoughtful discussions. On behalf of the planning committee, I would like to thank Dr. Christine Taylor, Dr. Linda Meyers, and all of the Food and Nutrition Board (FNB) and Institute of Medicine (IOM) staff who took the planning committee's input and created this successful meeting, especially Sandra Amamoo-Kakra, Heather Del Valle, and Gerri Kennedo.

The immediate next steps will be to make the presenters' slides available on the workshop website (http://www.iom.edu/driworkshop2007). Also, a preliminary draft of the workshop summary will be issued in the next several months. I am certain that the many themes we have heard throughout the workshop will be reflected in this summary of the presentations and discussions.

In terms of future action, the most important outcome will be to foster and guide the needed conversations among our government sponsors, the IOM leadership, and relevant stakeholders so that the next tasks can be identified and plans can be made to fulfill those tasks. The workshop participants have given us a wealth of information. We will digest and organize that information in a thoughtful and collaborative manner.

Presenter: Paul Coates

I am pleased to offer closing remarks on behalf of the U.S. and Canadian sponsors. Let me begin by saying that we owe a debt of thanks to those who devoted their time and intellect to the development of the DRIs. Furthermore, we wish to acknowledge the valuable role that the IOM has played, along with the study committees, in informing public health nutrition policy through these nutrient reference values. Both governments have benefited by having a sound scientific basis available for making appropriate nutrition policy decisions. Moreover, as you may know, the two governments have played a role in the development of these reference values. Overall, we can learn much from one another—and the hope is that these collaborations may serve as a starting point or even a model for similar developments around the world.

The U.S.–Canadian collaboration began with joint sponsorship of the development of the DRIs in the early 1990s. Both countries have DRI steering committees, and active joint discussions are a key component of our liaison activities. The overall effort has resulted in a series of important documents—the DRI volumes that have guided policy and informed dietary recommendations. More recently, the Canadian government sponsored the IOM preparation of a single-volume guide to the DRIs (IOM, 2006), published in English and French. The French version provides access to the 30 percent of Canadians for whom French is their first language.

Another collaborative effort is the recent project to synthesize and publish the entire set of research recommendations contained in the six volumes of the DRIs (IOM, 2007). The database associated with the project will soon be made completely accessible and highlights the knowledge needed to improve future DRI values.

This week's workshop on the DRI development process comes at the close of the decade-long DRI initiative and represents the culmination of several important collaborations. All groups represented here today have played a crucial role. In this respect, we have easily met the goals we hoped to accomplish in this meeting. Participants promoted a broad and critical evaluation of the current DRI model, and the meeting provided a locus for discussion of the lessons learned and the challenges we face in developing future DRI-related efforts.

Speaking on behalf of the workshop's sponsors, we were struck by the enormous value of the DRI initiative overall and by the remarkable candor of the meeting participants about their experiences in contributing to the initiative. Their willingness to offer this type of input is a measure of how important the DRI effort is and how committed the participants are to bringing the best information to bear on what ultimately supports our public health recommendations. Although consensus was not a component of this meeting's activities, we nonetheless obtained a great deal of food for thought.

We have learned a great deal from the experiences shared at this meeting, including some ongoing challenges associated with dietary recommendation issues. We now have a picture of what we need to consider as we move forward. Notable among the issues are the scope, the organizing framework, and the basis for revisiting or developing new DRI values, as well as the need for scientific guiding criteria and the incorporation of systematic review approaches in order to enhance transparency. On behalf of the government sponsors, we thank all involved for the astonishing amount of work that resulted in these fruitful discussions. The final product has certainly been worthy of their efforts.

References

Atkinson, S. A., and B. Koletzko. 2007. Determining life-stage groups and extrapolating nutrient intake values (NIVs). *Food and Nutrition Bulletin* 28(1 Suppl Int):S61–S77.

Ballew, C., D. Galuska, and C. Gillespie. 2001. High serum retinyl esters are not associated with reduced bone mineral density in the Third National Health and Nutrition Examination Survey, 1988–1994. *Journal of Bone and Mineral Research* 16:2306–2312.

Bernstein, A. L., and C. S. Lobitz. 1988. A clinical and electrophysiologic study of the treatment of painful diabetic neuropathies with pyridoxine abuse. In *Clinical and physiological applications of vitamin B6* (Current topics in nutrition and disease, Vol. 311), edited by J. Leklem and R. D. Reynolds. New York: Alan R. Liss.

Bischoff-Ferrari, H., W. C. Willett, J. B. Wong, E. Giovannucci, T. Dietrich, and B. Dawson-Hughes. 2005. Fracture prevention with vitamin D supplementation: A meta-analysis of randomized controlled trials. *Journal of the American Medical Association* 293: 2257–2264.

Blot, W. J., J. Y. Li, P. R. Taylor, W. Guo, S. Dawsey, G. Q. Wang, C. S. Yang, S. F. Zheng, M. Gail, G. Y. Li, Y. Yu, B. Q. Liu, J. Tangrea, Y. H. Sun, F. Liu, J. F. Fraumani, Jr., Y. H. Zhang, and B. Li. 1993. Nutrition intervention trials in Linxian, China: Supplementation with specific vitamin/mineral combinations, cancer incidence, and disease-specific mortality in the general population. *Journal of the National Cancer Institute* 85(18):1483–1492.

Dalton, K., and M. J. T. Dalton. 1987. Characteristics of pyridoxine overdose neuropathy syndrome. *Acta Neurologica Scandinavica* 76:8–11.

Davis, C. D., D. B. Milne, and F. H. Nielsen. 2000. Changes in dietary zinc and copper affect zinc-status indicators of postmenopausal women, notably, extracellular superoxide dismutase and amyloid precursor proteins. *American Journal of Clinical Nutrition* 71:781–788.

de Onis, M., A. W. Onyango, E. Borghi, C. Garza, H. Yang, and WHO Multicentre Growth Reference Study Group. 2006. Comparison of the World Health Organization (WHO) child growth standards and the National Center for Health Statistics/WHO international growth reference: Implications for child health programmes. *Public Health and Nutrition* 9(7):942–947.

de Onis, M., C. Garza, A. W. Onyango, and E. Borghi. 2007. Comparison of the WHO child growth standards and the CDC 2000 growth charts. *Journal of Nutrition* 137(1): 144–148.

DeGroot, L. J. 1966. Kinetic analysis of iodine metabolism. *Journal of Clinical Endocrinology and Metabolism* 26:149–173.

Devaney, B., P. Ziegler, S. Pac, V. Karwe, and S. I. Barr. 2004. Nutrient intakes of infants and toddlers. *Journal of the American Dietetic Association* 104(1 Suppl 1):s14–s21.

Duffield, A. J., C. D. Thomson, K. E. Hill, and S. Williams. 1999. An estimation of selenium requirements for New Zealanders. *American Journal of Clinical Nutrition* 70:896–903.

Etminan, M., J. M. FitzGerald, M. Gleave, and K. Chambers. 2005. Intake of selenium in the prevention of prostate cancer: A systematic review and meta-analysis. *Cancer Causes and Control* 16:1125–1131.

Fisher, D. A., and T. H. Oddie. 1969a. Thyroidal radioiodine clearance and thyroid iodine accumulation: Contrast between random daily variation and population data. *Journal of Clinical Endocrinology and Metabolism* 29:111–115.

Fisher, D. A., and T. H. Oddie. 1969b. Thyroid iodine content and turnover in euthyroid subjects: Validity of estimation of thyroid iodine accumulation from short-term clearance studies. *Journal of Clinical Endocrinology and Metabolism* 29:721–727.

Garza, C. 2006. New growth standards for the 21st century: A prescriptive approach. *Nutrition Reviews* 64(5 Pt 2):S55–S59; discussion S72–S91.

Garza, C., and D. L. Pelletier. 2007. Dietary guidelines past, present, and future. In *Nation's nutrition*, edited by E. Kennedy and R. Deckelbaum. Washington, DC: ILSI Press. P. 205.

Heinonen, O. P., and D. Albanes. 1994. The effect of vitamin E and beta-carotene on the incidence of lung cancer and other cancers in male smokers. *New England Journal of Medicine* 330(15):1029–1035.

Hunter, D. J., D. Spiegelman, H. O. Adami, L. Beeson, P. A. van den Brandt, A. R. Folsom, G. E. Fraser, R. A. Goldbohm, S. Graham, G. R. Howe, L. H. Kushi, J. R. Marshall, A. McDermott, A. B. Miller, F. E. Speizer, A. Wolk, S.-S. Yaun, and W. Willett. 1996. Cohort studies of fat intake and the risk of breast cancer—a pooled analysis. *New England Journal of Medicine* 334:356–361.

Ingenbleek, Y., and P. Malvaux. 1974. Iodine balance studies in protein-calorie malnutrition. *Archives of Disease in Childhood* 49:305–309.

IOM (Institute of Medicine). 1994. *How should the Recommended Dietary Allowances be revised?* Washington, DC: National Academy Press.

IOM. 1997. *Dietary Reference Intakes for calcium, phosphorus, magnesium, vitamin D, and fluoride.* Washington, DC: National Academy Press.

IOM. 1998. *A risk assessment model for establishing upper intake levels for nutrients.* Washington, DC: National Academy Press.

IOM. 2000a. *Dietary Reference Intakes: Applications in dietary assessment.* Washington, DC: National Academy Press.

IOM. 2000b. *Dietary Reference Intakes for vitamin C, vitamin E, selenium, and carotenoids.* Washington, DC: National Academy Press.

IOM. 2001. *Dietary Reference Intakes for vitamin A, vitamin K, arsenic, boron, chromium, copper, iodine, iron, manganese, molybdenum, nickel, silicon, vanadium, and zinc.* Washington, DC: National Academy Press.

IOM. 2002/2005. *Dietary Reference Intakes for energy, carbohydrate, fiber, fat, fatty acids, cholesterol, protein, and amino acids.* Washington, DC: The National Academies Press.

IOM. 2003a. *Dietary Reference Intakes: Applications in dietary planning.* Washington, DC: The National Academies Press.

IOM. 2003b. *Dietary Reference Intakes: Guiding principles for nutrition labeling and fortification.* Washington, DC: The National Academies Press.

IOM. 2006. *Dietary Reference Intakes: The essential guide to nutrient requirements.* Washington, DC: The National Academies Press.

IOM. 2007. *Dietary Reference Intakes: Research Synthesis Workshop Summary.* Washington, DC: The National Academies Press.

IPCS (International Programme on Chemical Safety). 2002. *Principles and methods for the assessment of risk from essential trace elements* (Environmental Health Criteria 228). Geneva, Switzerland: World Health Organization, http://www.inchem.org/documents/ehc/ehc/ehc228.htm.

IPCS. 2004. *IPCS risk assessment terminology* (Harmonization Project Document No. 1). Geneva, Switzerland: World Health Organization.

IPCS. 2005. *Chemical-specific adjustment factors for interspecies differences and human variability: Guidance document for use of data in dose/concentration–response assessment* (Harmonization Project Document No. 2). Geneva, Switzerland: World Health Organization.

Katamay, S. W., K. A. Esslinger, M. Vigneault, J. L. Johnston, B. A. Junkins, L. G. Robbins, I. V. Sirois, E. M. Jones-Mclean, A. F. Kennedy, M. A. Bush, D. Brulé, and C. Martineau. 2007. "Eating well with Canada's Food Guide (2007)": Development of the food intake pattern. *Nutrition Reviews* 65(4):155–166.

Kuczmarski, R. J., C. L. Ogden, L. M. Grummer-Strawn, K. M. Flegal, S. S. Guo, R. Wei, Z. Mei, L. R. Curtin, A. F. Roche, and C. L. Johnson. 2000. *CDC growth charts: United States, Advance Data No. 314.* Atlanta, GA: National Center for Health Statistics, http://www.cdc.gov/growthcharts.

Malvaux, P., C. Beckers, and M. De Visscher. 1969. Iodine balance studies in nongoitrous children and in adolescents on low iodine intake. *Journal of Clinical Endocrinology and Metabolism* 29:79–84.

Melhus, H., K. Michaelsson, A. Kindmark, R. Bergstrom, L. Holmberg, H. Mallmin, A. Wolk, and S. Ljunghall. 1998. Excessive dietary intake of vitamin A is associated with reduced bone mineral density and increased risk for hip fracture. *Annals of Internal Medicine* 29(10):770–778.

Milne, D. B., C. D. Davis, and F. H. Nielsen. 2001. Low dietary zinc alters indices of copper function and status in postmenopausal women. *Nutrition* 17:701–708.

Morris, R. C., Jr., A. Sebastian, A. Forman, M. Tanaka, and O. Schmidlin. 1999. Normotensive salt-sensitivity: Effects of race and dietary potassium. *Hypertension* 33(1):18–23.

Murphy, S. P., and S. I. Barr. 2007. Food guides reflect similarities and differences in dietary guidance in three countries (Japan, Canada, and the United States). *Nutrition Reviews* 65(4):141–148.

NRC (National Research Council). 1941. *Recommended Dietary Allowances.* Washington, DC: NRC.

NRC. 1958. *Recommended Dietary Allowances, revised 1958.* Report of the Food and Nutrition Board (Publication 3589). Washington, DC: NRC.

NRC. 1964. *Recommended Dietary Allowances, sixth revised edition.* Report of the Food and Nutrition Board (Publication 1146). Washington, DC: NRC.

NRC. 1968. *Recommended Dietary Allowances, seventh edition, 1968.* Report of the Food and Nutrition Board (Publication 1694). Washington, DC: NRC.

NRC. 1974. *Recommended Dietary Allowances, eighth revised edition, 1974.* Report of the Committee on Dietary Allowances and Committee on Interpretation of the Recommended Dietary Allowances. Washington, DC: NRC.

NRC. 1980. *Recommended Dietary Allowances, ninth edition.* Report of the Committee on Dietary Allowances, Food and Nutrition Board. Washington, DC: National Academy Press.

NRC. 1983. *Risk assessment in the federal government: Managing the process*. Committee on the Institutional Means for Assessment of Risks to Public Health, Commission on Life Sciences. Washington, DC: National Academy Press.

NRC. 1989a. *Recommended Dietary Allowances, 10th edition*. Report of the Subcommittee on the 10th Edition of the RDAs, Food and Nutrition Board. Washington, DC: National Academy Press.

NRC. 1989b. *Diet and health: Implications for reducing chronic disease risk*. Washington, DC: National Academy Press.

Office of the Surgeon General. 1988. *The Surgeon General's report on nutrition and health* (National Technical Information Service Order No. PB92-106947INZ). Washington, DC: U.S. Public Health Service.

Omenn, G. S., G. E. Goodman, M. D. Thornquist, J. Balmes, M. R. Cullen, A. Glass, J. P. Keogh, F. L. Meyskens, B. Valanis, J. H. Williams, S. Barnhart, and S. Hammar. 1996. Effects of a combination of beta carotene and vitamin A on lung cancer and cardiovascular disease. *New England Journal of Medicine* 334(18):1150–1155.

Philips, W. E., J. H. Mills, S. M. Charbonneau, L. Tryphonas, G. V. Hatina, Z. Zawidzka, F. R. Bryce, and I. C. Munro. 1978. Subacute toxicity of pyridoxine hydrochloride in the beagle dog. *Toxicology and Applied Pharmacology* 44:323–333.

Przyrembel, H. 2006. Discussion Paper 2: Uncertainty and adjustment. Annex 3 in *A model for establishing upper levels of intake for nutrients and related substances*. Report of a Joint FAO/WHO Technical Workshop on Nutrient Risk Assessment. Workshop held at World Health Organization Headquarters, Geneva, May 2–6, 2005. Geneva: World Health Organization. Pp. 123–156.

Renwick, A. G. 2006. Toxicology of micronutrients: Adverse effects and uncertainty. *Journal of Nutrition* 136:493S–501S.

Rosenzweig, A. 2007. Scanning the genome for coronary risk. *New England Journal of Medicine* 357(5):497–499.

Rothman, K. J., L. L. Moore, M. R. Singer, U. D. T. Nguyen, S. Mannino, and A. Milunsky. 1995. Teratogenicity of high vitamin A intake. *New England Journal of Medicine* 333(21):1369–1373.

SCF (Scientific Committee for Food). 1993. *Nutrient and energy intakes for the European Community* (Opinion expressed on December 11, 1992). Reports of the Scientific Committee for Food (Thirty-first series). Brussels, Luxembourg: Office for Official Publications of the European Communities. Pp. 1–248.

Sowers, M. R., and R. B. Wallace. 1990. Retinol, supplemental vitamin A and bone status. *Journal of Clinical Epidemiology* 43:693–699.

Subar, A. F., V. Kipnis, R. P. Troiano, D. Midthune, D. A. Schoeller, S. Bingham, C. O. Sharbaugh, J. Trabulsi, S. Runswick, R. Ballard-Barbash, J. Sunshine, and A. Schatzkin. 2003. Using intake biomarkers to evaluate the extent of dietary misreporting in a large sample of adults: The OPEN Study. *American Journal of Epidemiology* 158:1–13.

Suitor, C. W., and P. M. Gleason. 2002. Using Dietary Reference Intake-based methods to estimate the prevalence of inadequate nutrient intake among school-aged children. *Journal of the American Dietetic Association* 102(4):530–536.

Taubes, G. 2007. Why can't we trust much of what we hear about diet, health and behavior-related diseases? *New York Times Magazine*, September 16.

Taylor, P. R., Y. Qiao, S. M. Dawsey, L. L. Johnson, Z. Dong, and B. Yu. 2005. Total and cancer mortality following supplementation with multi-vitamins and minerals: Post-intervention follow-up of the General Population Trial in Linxian, China. *Gastroenterology* 128:A296.

Trabulsi, J., and D. A. Schoeller. 2001. Evaluation of dietary assessment instruments against doubly labeled water, a biomarker of habitual energy intake. *American Journal of Physiology–Endocrinology and Metabolism* 281:E891–E899.

Ulrich, C. M. 2007. Folate and cancer prevention: A closer look at a complex picture. *American Journal of Clinical Nutrition* 86:271–273.

Wang, C., M. Chung, A. Lichtenstein, E. Balk, B. Kupelnick, D. DeVine, A. Lawrence, and J. Lau. 2004. *Effects of omega-3 fatty acids on cardiovascular disease* (Evidence Report/ Technology Assessment No. 94). Rockville, MD: Agency for Healthcare Research and Quality.

Waterland, R. A., and R. L. Jirtle. 2003. Transposable elements: Targets for early nutritional effects on epigenetic gene regulation. *Molecular and Cellular Biology* 23(15): 5293–5300.

Waterland, R. A., D. C. Dolinoy, J. R. Lin, C. A. Smith, X. Shi, and K. Tahiliani. 2006. Maternal methyl supplements increase offspring DNA methylation at Axin Fused. *Genesis* 44(9):401–406.

WHO (World Health Organization). 1985. *Energy and protein requirements* (WHO Technical Report Series 724). Report of a joint FAO/WHO/UNU meeting. Geneva, Switzerland: WHO.

Wright, M. E., K. A. Lawson, S. J. Weinstein, P. Pietinen, P. R. Taylor, J. Virtamo, and D. Albanes. 2006. Higher baseline serum concentrations of vitamin E are associated with lower total and cause-specific mortality in the Alpha-Tocopherol, Beta-Carotene Cancer Prevention Study. *American Journal of Clinical Nutrition* 84:1200–1207.

Yadrick, M. K., M. A. Kenney, and E. A. Winterfeldt. 1989. Iron, copper, and zinc status: Response to supplementation with zinc or zinc and iron in adult females. *American Journal of Clinical Nutrition* 49:145–150.

Yang, G. Q., L. Z. Zhu, S. J. Liu, L. Z. Gu, P. C. Qian, J. H. Huang, and M. D. Lu. 1987. Human selenium requirements in China. In *Selenium in biology and medicine*, edited by G. F. Combs, Jr., O. A. Levander, J. E. Spallholz, and J. E. Oldfield. New York: Avi. Pp. 589–607.

Appendix A

Workshop Agenda and Background Materials

AGENDA

Institute of Medicine, Food and Nutrition Board Workshop

The Development of DRIs 1994–2004: Lessons Learned and New Challenges

The NAS Auditorium
2100 C Street, NW
Washington, DC

September 18–20, 2007

***** **DAY 1 – September 18** *****

7:30–8:30 am Registration

INTRODUCTION

8:30–9:00 **Welcome and Opening Remarks**
John Suttie, Chair, University of Wisconsin

9:00–9:30 **Overview of Current DRI Framework and Issues Raised**
Christine Taylor, Institute of Medicine

9:30–10:00 Break

SESSION 1
CONCEPTUAL FRAMEWORK FOR DRI DEVELOPMENT
Moderator: Stephanie Atkinson, McMaster University

10:00–10:40	**Current Framework for DRI Development: What Are the Pros and Cons?** *Robert M. Russell, Tufts University*
10:40–10:55	**Case Study: Applying the DRI Framework to Chronic Disease Endpoints** *Paula Trumbo, Center for Food Safety and Applied Nutrition, FDA*
10:55–11:10	**Case Study: Applying the DRI Framework to Non-Chronic Disease Endpoints** *Allison Yates, Agricultural Research Service, USDA*
11:10–11:50	***Discussion: Framework Pros/Cons; Case Studies*** **Co-Discussants:** *Patsy Brannon, Cornell University, and Alice H. Lichtenstein, Tufts University*
11:10–11:30	Discussion among co-discussants/presenters
11:30–11:50	Discussion open to all attendees
11:50 am–12:20 pm	**Two Perspectives: The DRI Framework** **Perspective I:** *George Beaton, University of Toronto* **Perspective II:** *Janet King, University of California, Berkeley and Davis*
12:20–12:30	***Question and Answer Session on Perspectives***
12:30–1:30	Lunch
1:30–1:45	**Evaluating Evidence for DRI Development: What Are the Issues in Applying Systematic Evidence-Based Review Approaches to DRI Development?** *Alice H. Lichtenstein, Tufts University*
1:45–2:15	**Risk Assessment: Is It a Relevant Organizing Structure?** *Elizabeth A. Yetley, Office of Dietary Supplements, NIH*
2:15–2:55	***Discussion: Systematic Evidence-Based Review; Risk Assessment*** **Discussant:** *Sanford Miller, University of Maryland*

2:15–2:35	Discussion among discussant/presenters
2:35–2:55	Discussion open to all attendees
2:55–3:15	Break
3:15–5:15	**Panel Discussion: In What Ways Could the Conceptual Framework Be Enhanced?** **Panel Members:** *Cutberto Garza, Boston College; Mary L'Abbé, Health Canada; Irwin Rosenberg, Tufts University; and Barbara Stoecker, Oklahoma State University*
3:15–3:35	Opening remarks from panel members
3:35–4:00	Cross-panel discussion
4:00–5:15	Discussion open to all attendees

***** DAY 2 – September 19 *****

SESSION 2

CRITERIA FOR SCIENTIFIC DECISION MAKING

Moderator: Robert M. Russell, Tufts University

8:30–8:50 am	**Selecting Endpoints: What Are the Issues and What Are the Options for Criteria?** *Irwin Rosenberg, Tufts University*
8:50–9:00	***Discussion***
9:00–9:15	**Dose–Response Data: Are There Options for Dealing with Limited Data?** *Susan Taylor Mayne, Yale School of Public Health*
9:15–9:30	***Discussion***
9:30–9:45	**What Are the Challenges in Addressing Extrapolation/Interpolation for Unstudied Groups?** *Stephanie A. Atkinson, McMaster University*
9:45–10:00	***Discussion***
10:00–10:30	Break
10:30–10:45	**What Are the Challenges in Addressing Adjustment for Data Uncertainty?**

	Hildegard Przyrembel, Federal Institute for Risk Assessment, Berlin, Germany
10:45–11:00	***Discussion***
11:00–11:20	**Estimating Dietary Intake: What Are the Implications for DRI Development?** *Amy Subar, National Cancer Institute, NIH*
11:20–11:30	***Discussion***
11:30–11:45	**Highlights of Other Important Issues: Physiological, Genomic, and Environmental Factors** *Cutberto Garza, Boston College*
11:45 am–12:00 pm	***Discussion***
12:00–1:00	Lunch

SESSION 3
GENERAL GUIDANCE FOR USERS OF DRIs
Moderator: Mary Bush, Health Canada

1:00–1:20 pm	**Overview: Issues Raised About General Guidance for Users** *Christine Taylor, Food and Nutrition Board, Institute of Medicine*
1:20–1:40	***Discussion*** **Discussant:** *Johanna Dwyer, Office of Dietary Supplements, NIH*
1:20–1:30	Discussion between discussant/presenter
1:30–1:40	Discussion open to all attendees
1:40–2:05	**Special Challenges: Planning and Assessing the Total Diet—What Are the Issues and What Are the Options for Enhanced Guidance?** *Suzanne Murphy, University of Hawaii* *Susan Barr, University of British Columbia*
2:05–2:30	***Discussion*** **Co-Discussants:** *Patricia Guenther, Center for Nutrition*

	Policy and Promotion, USDA, and Krista Esslinger, Health Canada
2:05–2:20	Discussion among co-discussants/presenters
2:20–2:30	Discussion open to all attendees
2:30–2:45	Break
2:45–3:10	**Special Challenges: What Are the Issues Related to a Framework for Individual-Level and Group-Level Applications?** *Valerie Tarasuk, University of Toronto*
3:10–3:30	***Discussion*** **Discussant:** *Gerard Dallal, Tufts University*
3:10–3:20	Discussion between discussant/presenter
3:20–3:30	Discussion open to all attendees
3:30–5:15	**Panel Discussion: In What Ways Could the Guidance for Users of DRIs Be Enhanced?** **Panel Members:** *Danielle Brulé, Health Canada; Mary Frances Picciano, Office of Dietary Supplements, NIH; William Rand, Tufts University School of Medicine; and Linda Van Horn, Northwestern University*
3:30–3:50	Opening remarks from panel members
3:50–4:15	Cross-panel discussion
4:15–5:15	Discussion open to all attendees

******* DAY 3 – September 20 *******

SESSION 4
LOOKING TO THE FUTURE PROCESS FOR DRI DEVELOPMENT

Moderator: Paul Coates, Office of Dietary Supplements, NIH

8:30–8:45 am	**Emerging Issues: What New Challenges Might the Future Hold?** *Catherine Woteki, Mars, Inc.*
8:45–8:55	***Discussion***
8:55–9:10	**Is There a Need to Enhance Transparency of the Decision-Making Process?** *Robert M. Russell, Tufts University*

9:10–9:20	***Discussion***
9:20–9:35	**IOM Overview of Options for Stakeholder Input** *Linda D. Meyers, Food and Nutrition Board, Institute of Medicine*
9:35–9:45	***Discussion***
9:45–10:00	**What Are the Criteria Options for Determining When to Update/Review Existing DRIs?** *John Suttie, Chair, University of Wisconsin*
10:00–10:10	***Discussion***
10:10–10:25	**What Are the Considerations in Specifying "New" Nutrient Substances for DRI Study?** *Peter Greenwald, National Cancer Institute, NIH*
10:25–10:35	***Discussion***
10:35–10:50	Break
10:50 am–12:45 pm	**Panel Discussion: Reflections on What We Have Heard About the Process of DRI Development** **Panel Members:** *Mary Bush, Health Canada; Jean-Pierre Habicht, Cornell University; Suzanne Harris, ILSI Research Foundation; Van Hubbard, Division of Nutrition Research Coordination, NIH; and Molly Kretsch, Agricultural Research Service, USDA*
10:50–11:15	Opening remarks from panel members
11:15–11:50	Cross-panel discussion
11:50–12:45	Discussion open to all attendees
12:45–1:15	**Summary and Closing Remarks** *Chair's Summary (John Suttie)* *Closing Remarks from Sponsor Representative (Paul Coates)* *Chair's Close of Workshop*

WORKSHOP BACKGROUND MATERIALS

The following background materials were made available via the IOM website to the general public and workshop participants for viewing prior to the workshop. The documents and the comments received can be accessed at www.iom.edu/driworkshop2007.

- IOM Publication: *How Should the Recommended Dietary Allowances Be Revised?* (1994)
- Tables: Comparisons of Outcomes/Approaches Among DRI Nutrients
- International Documents: Food and Agriculture Organization of the United Nations and the World Health Organization
- Paper: Risk Assessment: Is It a Relevant Organizing Structure? (Yetley)
- Paper: Selection of Endpoints for Determining EARs/AIs and ULs (Cheney)
- Paper: DRI Development Process; Issues Related to Extrapolation and Interpolation for Unstudied Groups (Atkinson)
- Paper: DRI Development Process; Issues Related to the Adjustment for Data Uncertainty (Przyrembel)
- Paper: DRI Development Process; Issues of Variability (Rand)
- Paper: Approximating Dose-Response in the Face of Limited Data (Mayne)
- Paper: Uses and Challenges in Applying the DRIs; U.S. Federal DRI Steering Committee
- Paper: Uses and Challenges in Applying the DRIs; Health Canada
- Paper: Uses and Challenges in Applying the DRIs; American Dietetic Association
- Paper: Uses and Challenges in Applying the DRIs; Dietitians of Canada

Appendix B

Workshop Presenters, Discussants, Panelists, and U.S./Canadian Sponsor Representatives

WORKSHOP PRESENTERS, DISCUSSANTS, AND PANELISTS

Stephanie Atkinson, Ph.D.
Department of Pediatrics
Faculty of Health Sciences
McMaster University
Hamilton, Ontario, Canada

Susan I. Barr, Ph.D., R.D., FACSM, FDC
Department of Food Nutrition and Health
University of British Columbia
Vancouver, British Columbia, Canada

George Beaton, Ph.D.
Department of Nutritional Science
Faculty of Medicine
University of Toronto
Toronto, Ontario, Canada

Patsy Brannon, Ph.D.
College of Human Ecology
Cornell University
Ithaca, NY

Danielle Brulé, Ph.D.
Research, Monitoring and Evaluation Division
Office of Nutrition Policy and Promotion
Health Products and Food Branch
Nepean, Ontario, Canada

Mary Bush, M.Sc., R.D.
Office of Nutrition Policy and Promotion
Health Products and Food Branch
Nepean, Ontario, Canada

Paul Coates, Ph.D.
Office of Dietary Supplements
National Institutes of Health
Rockville, MD

Gerard Dallal, Ph.D.
Biostatistics Unit
Jean Mayer USDA Human Nutrition Research Center on Aging
Tufts University
Boston, MA

Johanna Dwyer, D.Sc., R.D.
Office of Dietary Supplements
National Institutes of Health
Bethesda, MD

Krista Esslinger, M.Sc., R.D.
Health Canada
Vaudreuil-Dorion, Quebec, Canada

Cutberto Garza, M.D., Ph.D.
Boston College
Chestnut Hill, MA

Peter Greenwald, M.D., Dr.P.H.
Division of Cancer Prevention and Control
National Cancer Institute
National Institutes of Health
Bethesda, MD

Patricia M. Guenther, Ph.D., R.D.
USDA Center for Nutrition Policy and Promotion
Alexandria, VA

J. P. Habicht, Ph.D.
Division of Nutritional Sciences
Cornell University
Ithaca, NY

Suzanne Harris, Ph.D.
ILSI Research Foundation
Washington, DC

Van Hubbard, Ph.D., M.D.
Division of Nutrition Research Coordination
National Institutes of Health
Bethesda, MD

Janet King, Ph.D., R.D.
University of California, Berkeley and Davis
Children's Hospital Oakland Research Institute
Oakland, CA

Molly Kretsch, Ph.D.
USDA Agricultural Research Service
National Program Staff
Beltsville, MD

Mary L'Abbé, Ph.D.
Health Canada
Bureau of Nutritional Sciences
Ottawa, Ontario, Canada

Alice Lichtenstein, Ph.D.
Jean Mayer USDA Human Nutrition Research Center on Aging
Tufts University
Boston, MA

Susan Taylor Mayne, Ph.D., F.A.C.E.
Division of Chronic Disease Epidemiology
Yale School of Public Health
New Haven, CT

Linda D. Meyers, Ph.D.
Food and Nutrition Board
Institute of Medicine

Sanford Miller, Ph.D.
Center for Food, Nutrition, and Agriculture Policy
University of Maryland
College Park, MD

Suzanne P. Murphy, Ph.D., R.D.
Nutrition Support Shared Resource
Cancer Research Center of Hawaii
University of Hawaii
Honolulu, HI

Mary Frances Picciano, Ph.D.
Office of Dietary Supplements
National Institutes of Health
Bethesda, MD

Hildegard Przyrembel, M.D.
Federal Institute for Risk Assessment
Berlin, Germany

William M. Rand, Ph.D.
Department of Public Health and Family Medicine
Tufts University School of Medicine
Boston, MA

Irwin H. Rosenberg, M.D.
Friedman School of Nutrition Science and Policy
Tufts University
Boston, MA

Robert M. Russell, M.D.
Jean Mayer USDA Human Nutrition Research Center on Aging
Tufts University
Boston, MA

Barbara Stoecker, Ph.D.
Oklahoma State University
Stillwater, OK

Amy Subar, Ph.D.
National Cancer Institute
National Institutes of Health
Rockville, MD

John W. Suttie, Ph.D.
Department of Biochemistry
University of Wisconsin
Madison, WI

Valerie Tarasuk, Ph.D.
Department of Nutritional Sciences
University of Toronto
Toronto, Ontario, Canada

Paula Trumbo, Ph.D.
U.S. Food and Drug Administration
Center for Food Safety and Applied Nutrition
College Park, MD

Linda Van Horn, Ph.D., R.D.
Associate Dean for Faculty Development
Northwestern University
Feinberg School of Medicine
Chicago, IL

Catherine E. Woteki, Ph.D., R.D.
Mars, Inc.
McLean, VA

Allison Yates, Ph.D.
Beltsville Human Nutrition Center
Agricultural Research Service
Beltsville, MD

Elizabeth A. Yetley, Ph.D.
Office of Dietary Supplements
National Institutes of Health
Bethesda, MD

U.S./CANADIAN SPONSOR REPRESENTATIVES

U.S. Department of Health and Human Services

*Office of Disease Prevention and Health Promotion**
Kathryn McMurry, M.S.
Administration on Aging
Jean Lloyd, Ph.D.
Centers for Disease Control and Prevention
Division of Nutrition and Physical Activity
Joel Kimmons, Ph.D.
National Center for Health Statistics*
Clifford Johnson, M.S.P.H.
Food and Drug Administration
Center for Food Safety and Applied Nutrition*
Paula Trumbo, Ph.D.
National Institutes of Health
Division of Nutrition Research Coordination*
Van Hubbard, M.D., Ph.D., and Pamela Starke-Reed, Ph.D.
National Cancer Institute*
John Milner, Ph.D.
National Heart, Lung, and Blood Institute
Darla Danford, Sc.D.
Office of Dietary Supplements*
Paul Coates, Ph.D., Rebecca Costello, Ph.D., and Elizabeth Yetley, Ph.D.

U.S. Department of Agriculture

Agricultural Research Service
Headquarters, National Human Nutrition Program*
Molly Kretsch, Ph.D., and David Klurfeld, Ph.D.
Beltsville Human Nutrition Research Center*
Allison Yates, Ph.D.
Center for Nutrition Policy and Promotion
Patricia Guenther, Ph.D.

*DRI Review Workshop Sponsors.

Cooperative State Research, Extension, and Education System
Susan Welsh, Ph.D.
Economic Research Service
Joanne Guthrie, Ph.D., and Elizabeth Frazao, Ph.D.
Food and Nutrition Service
Jay Hirschman, M.P.H., and Anita Singh, Ph.D.

U.S. Department of Defense

U.S. Army Research Institute of Environmental Medicine
Col. Karl Friedl, Ph.D.

Health Canada

*Food Directorate, Bureau of Nutritional Sciences**
Mary L'Abbé, Ph.D., and Peter Fischer, Ph.D.
*Office of Nutrition Policy and Promotion**
Mary Bush, M.Sc., R.D., Danielle Brulé, Ph.D., Margaret Cheney, Ph.D., and Krista Esslinger, M.Sc.

Canadian Institutes of Health Research

*Institute of Nutrition, Metabolism and Diabetes**
Hasan Hutchinson, Ph.D., N.D.

U.S./CANADIAN GOVERNMENT SPONSOR REPRESENTATIVES TO THE WORKSHOP PLANNING COMMITTEE

Danielle Brulé, Ph.D.
Peter Fischer, Ph.D.
Hasan Hutchinson, Ph.D.
Clifford Johnson, M.S.P.H.
David Klurfeld, Ph.D.
Molly Kretsch, Ph.D.
Kathryn McMurry, M.S.
John Milner, Ph.D.
Pamela Starke-Reed, Ph.D.
Paula Trumbo, Ph.D.
Elizabeth Yetley, Ph.D.

---

*DRI Review Workshop Sponsors.

Appendix C

Brief List of Reoccurring Workshop Discussions[1]

Crosscutting Topics	
Transparency	• Transparency was acknowledged as an important component of enhancing future Dietary Reference Intakes (DRIs). • Some suggested that additional efforts to document decisions would be a major step toward more transparency.
Precision and uncertainty surrounding reference values	• Concern was expressed that, as presented, the DRI values appear as "very certain" numbers, or at least that the level of confidence for all values is the same. • Some suggested additional text or a specific risk characterization step to clarify the confidence in or precision of the established reference values; other options were also discussed, which included use of asterisks, a numeric grading system, or expressing values as a range.

[1]This list, prepared by the rapporteurs and based on the workshop discussions, reflects suggestions made by presenters, discussants, and other workshop participants in relation to the workshop's focus. It was prepared for the convenience of the reader. It should not be construed as representing recommendations or consensus statements, nor is it reflective of all topics nor the entire breadth of the discussions.

Criteria for updating current DRIs	• A strategy for updating DRIs was identified by many as an urgent matter. • Some suggested that several venues may operate simultaneously and that relevant criteria need to be established; resources were acknowledged as a stumbling block.
Failure to establish reference values: No decision is not an option	• An educated estimate from scientists was recognized as a better alternative to not developing a reference value: A value derived from scientific judgment offers a basis for government managers who must act regardless of the existence of a value. • Interest was expressed in determining ways to specify relative uncertainty surrounding reference values and ways to identify controversies and concerns.
Stakeholder input	• Considerable opportunities for input were noted. • The rigor and independence of the Institute of Medicine (IOM) process were recognized.
Conceptual Underpinnings	
Uses and purpose of DRIs	• The overall goal of planning and assessing for groups and individuals was affirmed. • Concern was expressed that the endpoints selected cause confusion about what the DRIs are intended to accomplish.
Values expressed	• Estimated Average Requirements (EARs) and Tolerable Upper Intake Levels (ULs) have been useful. • Recommended Dietary Allowances (RDAs) were noted as useful to many, but were also characterized as arbitrary, misused, and more appropriately established using situation-specific criteria. • Adequate Intakes (AIs) were controversial and a source of confusion. Some saw no other option; some preferred establishing an EAR with an indication of uncertainty; some suggested that AIs can be relevant to use with endpoints based on chronic disease. • Some commented that consideration should be given to whether the DRI process should focus on a core set of "numbers" needed versus providing reference values for all applications.

Nature of endpoints	• Challenges in setting DRIs based on chronic disease endpoints were acknowledged. • Some suggested that chronic disease endpoints (with more data, better elucidation of confounders, and newer techniques for approximating dose–response) can be placed appropriately within the spectrum of nutritional effects; others suggested that standards for chronic disease need to be addressed separately from those for prevention of deficiency. • Concerns were expressed about providing multiple endpoints for a single age/gender group because it would be confusing and undermine the purpose of DRIs.
Nutrient substances appropriate for DRI consideration	• Many expressed interest in continuing to move beyond essential nutrients; some indicated that nonessential substances may require a different approach; some expressed interest in limiting DRIs to essential nutrients.
Road Map for DRI Development	
Systematic evidence-based reviews	• Such reviews were acknowledged as useful and relevant if the appropriate questions are articulated for the review. • They were also recognized as not relevant for all aspects of the DRI process; there was particular interest in ensuring that scientific judgment regarding the values to be established remains within the domain of the subject matter experts. • Concern was expressed about costs and time involved.
Risk assessment as an organizing scheme	• Risk assessment was acknowledged as relevant to the DRI process and as helpful in delineating roles and enhancing transparency and usability of outcomes. • The need to adapt the approach specifically for use with nutrient substances was recognized.
Scientific Decision-Making Criteria	
Selection of endpoints	• The need for specific criteria was acknowledged.
Approximation of dose–response relationship with limited data	• Useful techniques—both statistical and biological—have emerged and can be applied. • Concerns were expressed about statistical approaches when dealing with chronic disease endpoints and about biological techniques relative to the ability to link to human outcomes.

Extrapolation/ scaling	• This methodology was identified as necessary given the current state of datasets, but needs a stronger scientific foundation and consistent application.
Adjustment of data uncertainty	• Such adjustments were considered relevant to DRI development, but need a systematic approach.
Guidance for Users	
Organizational framework: 2×2 table	• Some indicated the table's utility as a basic starting point to address DRI applications; some indicated that it is overly simplistic and does not match real-world applications; some suggested it is too rigid.
Distinction between individual and group applications	• For some, the distinction is unclear; some indicated that the interface between individuals and groups has been missed; others suggested it is not a useful distinction if it causes a focus on applying the numbers rather than the underlying concepts.
General guidance versus specific guidance	• The diverse needs of users were acknowledged. • Some suggested the need to separate general guidance from specific guidance: Guidance for specific applications should be done on a case-by-case basis via separate reports. • There was interest in helping practitioners to obtain training and tools appropriate for their particular applications.

Appendix D

Schematic Used By Workshop Planning Committee: Activities Associated with DRI Development[1]

[1]Highlighted components were identified as the subjects of the September 2007 Institute of Medicine workshop on Dietary Reference Intake (DRI) development.

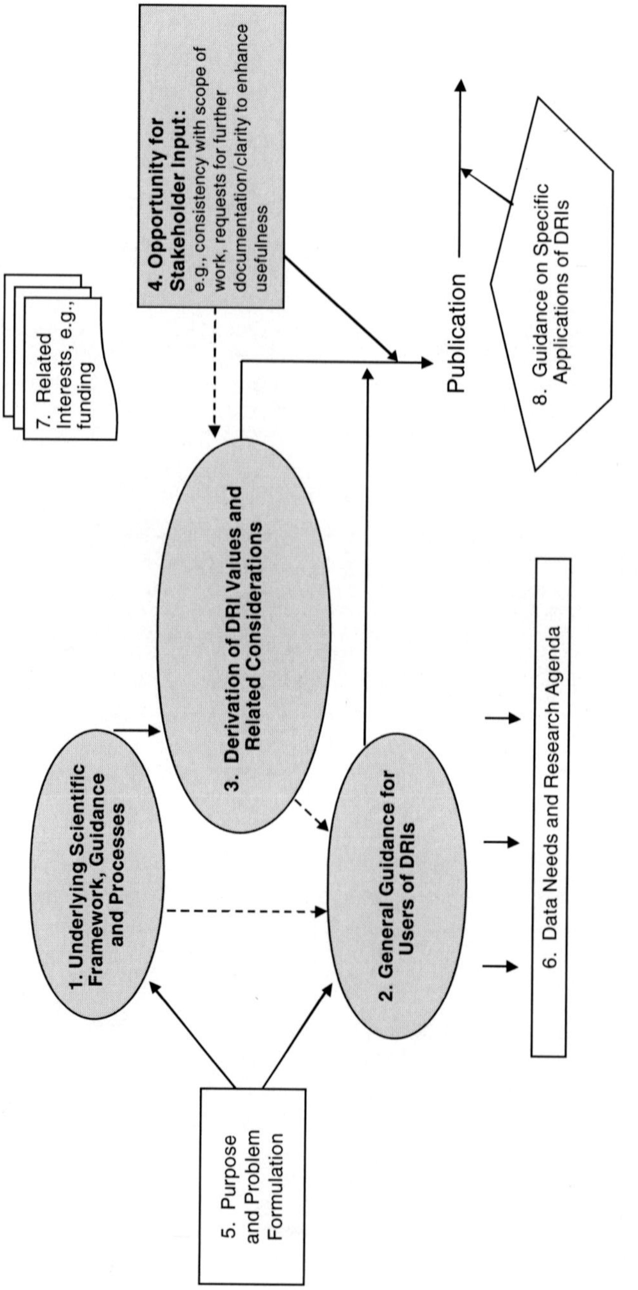
1. Underlying Scientific Framework, Guidance and Processes
2. General Guidance for Users of DRIs
3. Derivation of DRI Values and Related Considerations
4. Opportunity for Stakeholder Input:
e.g., consistency with scope of work, requests for further documentation/clarity to enhance usefulness
5. Purpose and Problem Formulation
6. Data Needs and Research Agenda
7. Related Interests, e.g., funding
8. Guidance on Specific Applications of DRIs
Publication